Object Detection and Image Classification

Object Detection and Image Classification

Guest Editors

Patrick Wong
Yifan Zhao

MDPI

Basel • Beijing • Wuhan • Barcelona • Belgrade • Novi Sad • Cluj • Manchester

Guest Editors
Patrick Wong
School of Computing and Communications
The Open University
Milton Keynes
UK

Yifan Zhao
Centre for Life-Cycle Engineering and Management
Cranfield University
Cranfield
UK

Editorial Office
MDPI AG
Grosspeteranlage 5
4052 Basel, Switzerland

This is a reprint of the Special Issue, published open access by the journal *Applied Sciences* (ISSN 2076-3417), freely accessible at: https://www.mdpi.com/journal/applsci/special_issues/35EU8135IZ.

For citation purposes, cite each article independently as indicated on the article page online and as indicated below:

Lastname, A.A.; Lastname, B.B. Article Title. *Journal Name* **Year**, *Volume Number*, Page Range.

ISBN 978-3-7258-5545-2 (Hbk)
ISBN 978-3-7258-5546-9 (PDF)
https://doi.org/10.3390/books978-3-7258-5546-9

Contents

applied sciences

MDPI

Editorial

Object Detection and Classification with Limited Training Data

Patrick Wong [1,*] and Yifan Zhao [2]

[1] School of Computing and Communications, Faculty of Science, Technology, Engineering and Mathematcis, Open University, Walton Hall, Milton Keynes MK7 6AA, UK
[2] Faculty of Engineering and Applied Sciences, Cranfield University, College Road, Cranfield, Bedfordshire MK43 0AL, UK; yifan.zhao@cranfield.ac.uk
* Correspondence: patrick.wong@open.ac.uk

1. Introduction

Since the rise of deep learning around a decade ago, the field of object detection and classification using a convolutional neural network (CNN) and its variants has grown exponentially. This technology has been applied in domains such as the manufacturing, construction, surveillance and monitoring, sports, transports, and medical sectors. An attractive quality of CNNs is their ability to take images in their raw form without the traditional feature extraction step of reducing input dimensionality. However, a significantly larger amount of training samples is required for CNNs to extract features and automatically classify objects. Labelling a large amount of data samples is costly and time-consuming, and as a result, the availability of training data is limited, particularly in domains outside of the science sectors. The labelling quality can also impact classification performance and reliability as wrong or erratic labelling can lead to poor or biassed classification performance. The availability of quality data has become a bottleneck, and it hinders CNNs' use in wider applications.

2. Learning with Limited Training Data

To address this bottleneck, the research community has developed various strategies to reduce the reliance on large amounts of training data. Data augmentation and data generation is one approach to produce more data from existing data. While data augmentation produces more variations of data by transforming the existing data in different ways, data generation achieves this by using generative models such as Generative Adversarial Networks [1]. However, as the produced data are derived from the original data, they do not often contain the extra features found in the original data samples.

Transfer learning [2] is a different approach to address the data scarcity bottleneck. It re-trains some parts, often the outer layers, of an existing model that was designed for classifying different but related objects. As the inner layers of the existing model have already learnt to recognise certain common features between old and new objects, less data are required to train the model to classify new objects.

Based on a similar idea, n-shot learning [3] aims to learn how to recognize new objects with just one or a few samples. It typically involves a meta learning process which aims to help a model quickly learn new tasks by training it on a variety of tasks with a few data samples. Training a model to predict the similarity between data points helps it generalise well with new tasks. Models also commonly use an embedding method in which the model learns to map inputs into a space where similar items are situated nearby. The model is often fine-tuned for a specific task with a small dataset through transfer learning.

3. Conclusions

With advances in data generation and transfer and meta learning, it is becoming feasible to train a model with limited labelled data. This will enable object detection to be applied to domains in which this was previously impossible due to a lack of labelled data. However, regarding computer vision, current object detection applications are still merely focused on detecting specific objects in images or videos rather than attempting to understand the holistic view of the image that is being represented. The first step of understanding an image is knowing what objects are in the image. However, understanding the relationships and interactions between these objects and predicting their future actions and behaviours, for example, are also necessary to reveal the deeper context of an image. Many more studies are needed to bring computer vision closer to human vision.

Author Contributions: Writing—original draft preparation, P.W.; writing—review and editing, Y.Z. All authors have read and agreed to the published version of the manuscript.

Conflicts of Interest: The authors declare no conflicts of interest.

References

1. Goyal, M.; Mahmoud, Q.H. A Systematic Review of Synthetic Data Generation Techniques Using Generative AI. *Electronics* **2024**, *13*, 3509. [CrossRef]
2. Zhuang, F.; Qi, Z.; Duan, K.; Xi, D.; Zhu, Y.; Zhu, H.; Xiong, H.; He, Q. *A Comprehensive Survey on Transfer Learning*; IEEE: Piscataway, NJ, USA, 2019. [CrossRef]
3. Parnami, A.; Lee, M. Learning from Few Examples: A Summary of Approaches to Few-Shot Learning. *arXiv* **2022**, arXiv:2203.04291.

applied sciences

MDPI

Article

Vision AI System Development for Improved Productivity in Challenging Industrial Environments: A Sustainable and Efficient Approach

Changmo Yang [1,2], JinSeok Kim [2], DongWeon Kang [2] and Doo-Seop Eom [1,*]

1 Department of Electrical and Computer Engineering, Korea University, 145 Anam-ro, Seongbuk-gu, Seoul 02841, Republic of Korea; cmyang83@gmail.com
2 Hyundai Motor Company, 37 Cheoldobangmulgwan-ro, Uiwang 16088, Republic of Korea; cmyang@hyundai.com (C.Y.)
* Correspondence: eomds@korea.ac.kr

Abstract: This study presents a development plan for a vision AI system to enhance productivity in industrial environments, where environmental control is challenging, by using AI technology. An image pre-processing algorithm was developed using a mobile robot that can operate in complex environments alongside workers to obtain high-quality learning and inspection images. Additionally, the proposed architecture for sustainable AI system development included cropping the inspection part images to minimize the technology development time, investment costs, and the reuse of images. The algorithm was retrained using mixed learning data to maintain and improve its performance in industrial fields. This AI system development architecture effectively addresses the challenges faced in applying AI technology at industrial sites and was demonstrated through experimentation and application.

Keywords: vision AI; industrial sites; AI technology; image pre-processing; mobile robot; sustainable AI system

1. Introduction

In recent years, extensive research has been conducted on the development of machine vision technology for product quality and work-ability improvement in industrial fields [1,2]. The rapid development of deep learning technology has enabled us to perform quality inspections on products with a diversity and complexity that was difficult to achieve in the past. However, the implementation of deep learning technology enables quality inspection primarily on small unit parts that can obtain relatively high-quality inspection images in an extremely limited environment. However, it is challenging to implement an inspection system using deep learning technology in an industrial environment where it is difficult to obtain high-quality images due to the structural diversity and frequent environmental changes, such as the assembly process of automobile manufacturing plants. Over 70% of automobile manufacturing plants use automated systems for assembly. However, parts such as wiring and connectors, which are difficult to automate due to their flexibility and versatility, still depend on manual assembly. Due to the structural characteristics of automobiles, during the assembly of parts, the next assembly part covers the previous assembly part, making it impossible to detect assembly defects through visual inspection using the human eye in the next process. If defects and omissions are detected through electrical inspection after the vehicle is assembled, the parts must then be disassembled in the reverse order to rectify these assembly defects, which involves considerable time and cost. Additionally, defective assembly of fixing clips, bolts, nuts, and so on cannot be detected electronically due to field claims such as vibration and noise produced while driving a vehicle. Recently, there has been a shift in the assembly process from the conveyor method to a cellular production method that is efficient for the small-volume production of multiple vehicles and options. The cellular method enables the assembly of up to five times more parts than conventional conveyors within a single process. This has increased the probability of parts assembly omission as well as erroneous assembly by workers, and has also increased the cost of poor assembly quality. Therefore, it is crucial to develop a new visual inspection system technology for real-time image acquisition and assembly defect detection after manual assembly by workers in an environment where it is difficult to obtain high-quality images in real-time. Two factors must be considered for real-time visual inspection in the manual assembly processes. Firstly, images must be obtained in real-time by collaborating with workers. Secondly, a visual inspection algorithm must be developed to detect assembly defects using the obtained images.

1.1. Image Acquisition Device

It is difficult to control lighting and environmental changes during manual assembly processes in the automobile industry, unlike small parts inspections. Furthermore, the mixed production of multiple vehicle models presents difficulties in image acquisition due to frequent changes in the inspection items. Therefore, highly mobile robots and manipulators with excellent flexibility are necessary for image acquisition in automobile manual assembly processes. Additionally, a device with high safety standards, obstacle avoidance, and excellent mobility that can be operated by workers must be developed. Various commercial devices have been implemented for image acquisition, but it is difficult to find a suitable device for manual assembly processes in the automobile industry. Moreover, it is challenging to acquire images inside the vehicle with a fixed camera. Although a 360-degree camera can be installed inside the vehicle before assembly to capture images, it is not suitable for inspection purposes due to low image resolution. Wearable glasses produce low-quality images due to shaking, and drones are not applicable due to safety and noise issues. A device must be developed that can collaborate flexibly with workers in automobile manual assembly processes and acquire high-quality image data through stability, obstacle avoidance, and excellent mobility.

1.2. Vision Inspection Algorithm Development

When a new car is introduced into a car production plant, the car body, color, and parts are changed. For approximately 100 days after the production of the vehicle, there are several defective parts assemblies due to the inexperience of the workers. After 100 days, the skill level of the operator improves, and the number of assembly defects is drastically reduced. Therefore, an assembly defect inspection system must be established at the initial stage of the production of new cars. However, the effect of implementing the existing rule-based visual inspection system is insufficient since an engineer requires at least six months to develop the visual inspection algorithm corresponding to the parts to be assembled in a new car. Several attempts are being made to apply deep learning technology to reduce the development period of the rule-based algorithm [3]. However, deep learning vision technology also requires high-quality learning data to develop new car assembly inspection algorithms. It takes more than 100 days to acquire the normal and defective data required for deep learning algorithm learning, due to which the visual inspection of assembly parts in the early stages of new car production is impossible, similar to the existing rule-based vision system. In this study, we propose a solution for the manual process, such as car assembly in the production plant, where it is difficult to implement the existing visual inspection system due to environmental changes and inspection item modifications. Our proposed solution includes a mobile robot suitable for image acquisition and a method for developing AI algorithms that can reduce the development period. We also demonstrate the performance of our proposed solution.

2. Related Works

Deep learning technology has been reported to outperform conventional rule-based visual inspection systems in detecting various types of assembly defects and conducting quality inspections in complex industrial settings [4–6]. However, high-quality training images are required to improve the performance of deep learning-based inspection systems. Most studies conducted on deep learning technology have been implemented on small-scale inspection targets with limited variation under controlled lighting and environmental conditions. In several industrial settings, it is difficult to obtain high-quality training images due to uncontrolled lighting and environmental conditions, along with frequent changes in the inspection targets, making it difficult to acquire sufficient training data. Previous studies have attempted to implement image acquisition and deep learning inspection under conditions similar to those of actual industrial sites. In Wang's study, assembly workers directly used wearable lenses and headsets to acquire images. The time required to capture the part images was set by using the worker's position and gaze information [7]. The evaluation results demonstrated that the system accuracy was low, at 85%. This was because even in a laboratory environment where the lighting conditions were consistent, there were image variations based on the distance and angle at which the worker captured the image. In Mazzetto's study, deep learning technology was implemented to inspect the surface treatment quality of automobile assembly parts. After sufficient training data were obtained, deep learning technology showed superior inspection performance when compared to the existing rule-based visual inspection method [3]. However, the inspection algorithm was only limited to the surface inspection of brake pedal parts that did not change when a new vehicle was released, and it was developed only after obtaining sufficient training data. Research is being conducted to address the challenge of obtaining sufficient data in industrial settings. The data acquired in actual industrial settings are small in quantity, but there is a significant difference in the OK and NG ratios. In the case of manual assembly processes for automobiles, data acquisition can be performed with an OK rate of 99% and an NG rate of 1%. Therefore, NG items must be created arbitrarily to acquire sufficient training data within a short period. This involves substantial time and cost. The one-class neural-network method, a type of semi-supervised learning, has been proposed to address this issue. This method learns using only a small number of normal images and detects samples that differ from the normal samples as outliers [8]. However, uncontrollable environments such as lighting can cause severe image variations due to image exposure. In the case of automobile parts comprising flexible cables and connectors, the position of the cable varies even in normal images, along with the position of the surrounding parts. Therefore, it is difficult to determine the boundary between bad and normal images. Even if an appropriate boundary is set, it frequently changes due to the rapidly changing environmental conditions, resulting in decreased inspection accuracy and increased maintenance costs. Therefore, a system must be developed that can acquire and inspect images under conditions where it is difficult to secure learning data of sufficient quality due to frequent changes in the inspection parts at industrial sites. Additionally, the high time and cost requirements for repeatedly maintaining detection algorithms due to frequent changes in the inspection items make it difficult to implement deep learning technology. In actual industrial sites, developing and implementing a deep learning inspection system can be challenging owing to the high cost required for the development and maintenance of the inspection system. In this study, we present a development methodology for a deep learning visual inspection system that can be implemented in an actual industrial field. First, image acquisition and pre-processing techniques using mobile robots acquire high-quality image data required to train the deep learning algorithms. Second, we propose a method to improve the performance of deep learning algorithms by using a small amount of data that can be acquired within a short period of time in actual industrial sites. Third, we propose a deep learning algorithm for the re-learning method to respond to frequent changes in the inspection parts and environmental conditions in industrial sites and to maintain the detection performance. Finally, the proposed technology is demonstrated through empirical evaluation.

3. Proposal Method

In this study, we propose a methodology to develop visual AI technology that enables the effective inspection of assembly defects in automobile production from the early stages, using mobile robot technology for image acquisition. We focus on the development process and system architecture to create a visual AI inspection system capable of accurately detecting incorrect parts assembly by workers in the manual assembly process of an automobile manufacturing plant, as illustrated in Figure 1.

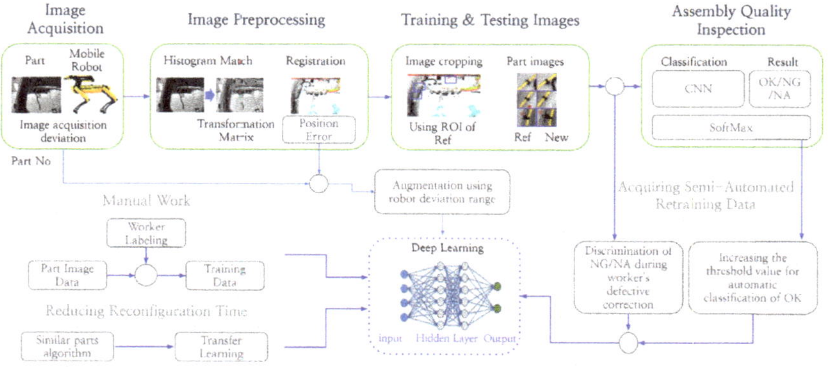

Figure 1. Flowchart of the proposed AI system for industrial site inspection.

We propose a new system to overcome the difficulties faced in implementing visual AI inspection in industrial settings, as depicted in Figure 2. Based on a standard visual AI inspection framework, this enhanced system additionally compensates for the variation in the images acquired by mobile robots due to changes in the industrial environment (①), and utilizes this information to crop specific component images (②) to generate optimized images that are crucial for assessing the assembly quality. Since automobile production facilities use more than 100 different components, each assembled in unique ways, the efficiency of visual AI inspections must be enhanced by individually applying tailored algorithms for each component (③), rather than employing a single generic algorithm. This methodology enables the straightforward reuse of algorithms for similar parts upon the introduction of new vehicle models or the application of simple transfer learning, thereby enhancing the operational management efficiency. From an operational perspective in industrial settings, we propose introducing an 'inspection error (NA)' category in deep learning image classification (④) beyond the conventional OK/NG criteria to mitigate productivity loss due to pseudo-defects and prevent the leakage of assembly defects. This raises the benchmark and requires operator verification when the criteria are not met, enabling the system to be operational even before sufficient training data have been accumulated. The inspection error images selected by the operators (⑤) can then be used as evaluation data for algorithm retraining, which streamlines the algorithm assessment and improvement process. A separate system must be established to maintain and manage the algorithm performance in case environmental changes in the industrial site cause variations in the mobile robot's positioning, potentially degrading the performance of pre-set detection algorithms (⑥–⑫), as suggested in Figure 1. This involves calculating the variance in the images during the cropping process, extracting a T-Matrix (⑦) through feature detection and matching (⑥), and storing the T-Matrix for each robot position (⑧) to calculate the range of variance. By employing this approach in deep learning algorithms, it enhances the algorithm performance through image augmentation techniques (⑨), utilizing the T-Matrix to define the range of variance within which the robot can acquire images, thus improving the detection algorithm performance. This strategy maintains and enhances the algorithm performance (⑩), enables the development of new algorithms through retraining (⑪), and uses evaluation data created by operator selection processes to compare and evaluate the performance of old and new algorithms, facilitating algorithm replacement if necessary (⑫). This approach ensures the continuous improvement in and maintenance of the algorithm performance.

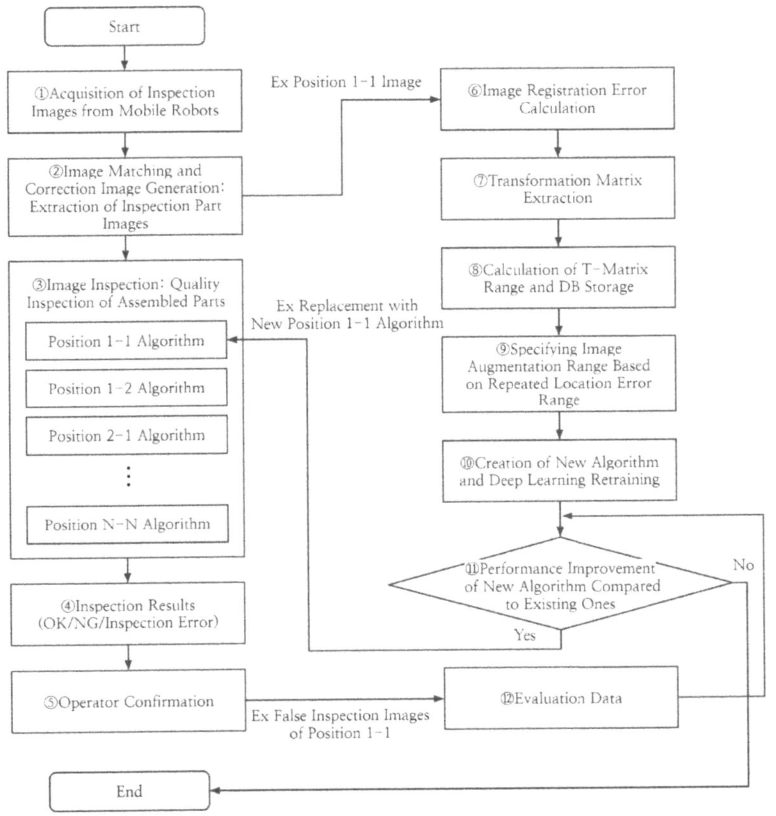

Figure 2. Integrated vision AI inspection system flowchart for quality testing in industrial environments.

4. Detailed Proposed Technology and Test Results

4.1. Acquisition of High-Quality Deep Learning and Inspection Data

In environments such as the manual assembly process in car production factories, which are narrow and complex, and where obstacles like bolts and nuts exist, a visual acquisition device must be developed that can acquire images in real-time and in the same space as the operator. In this study, we utilized a Boston Dynamics' (Waltham, MA, USA) four-legged robot, called SPOT, which has a relatively small body and the ability to move through narrow spaces with its four-legged walking system, as well as the self-SLAM technology that enables it to avoid obstacles. Additionally, the 4K camera attached to the SPOT package's seven-axis robot arm enables the easy acquisition of the part images [9]. However, due to the characteristics of the four-legged walking system, the repeated positioning accuracy exhibits a deviation of more than ±200 mm from the body base; there is also deviation when acquiring images using the camera attached to the arm. This is a common problem with all mobile robots used in industrial sites, and it can degrade the quality of the data, thereby affecting the performance of AI inspection. Unlike stationary robots that employ positional constraints, mobile robots utilizing methods such as visual SLAM can experience location errors ranging from 10 cm to 1 m. This variation in the positioning accuracy further complicates data acquisition and can significantly impact the effectiveness of AI-based inspections. Therefore, a solution must be developed to address this issue.

4.1.1. Landmark (Fiducial Mark) Centering Technique for Improving the Repeat Positioning Accuracy of SPOT

The SPOT robot uses five ToF cameras on its body for visual SLAM-based position movement. However, the reference vehicle moves using an AGV or conveyor for component imaging, causing position dispersion. Furthermore, the characteristics of quadrupedal walking result in poor repeat positioning accuracy with errors of over ±200 mm, as shown in Figure 3a. To address this problem, short-range communication devices, such as UWB technology, have been used to improve the repeat positioning accuracy in the industrial field [10,11]. However, this method incurs additional costs for installing infrastructure such as UWB transceivers in the surrounding environment, as well as AGV or vehicle attachment and removal of UWB. The location accuracy may also be reduced in environments such as car factory structures that can cause wireless signal fading. In this study, a landmark (Fiducial Mark) was attached to the AGV to consider the characteristics of the industrial field and minimize investment costs, which serves as the reference for position movement. A vertical reference point was specified between the SPOT body and the F-Mark to align the body accurately. This method does not require additional modification or cost even when the process changes or when new vehicles are introduced, as only the F-Mark must be attached without requiring additional infrastructure installation. Using this proposed method, the SPOT robot utilizes its front-facing camera to recognize the size of the attached F-Mark and measures its size and angle upon reaching the inspection site, as depicted in Figure 3b. By centering the SPOT body to be perpendicular to the F-Mark and adjusting the pre-set distance values, the positional error was reduced from ±200 mm to as low as ±14 mm. However, even if the error of the SPOT body is small, the positional error of the camera at the end of the arm that acquires the image accumulates based on the arm pose, causing a large deviation in the acquired image.

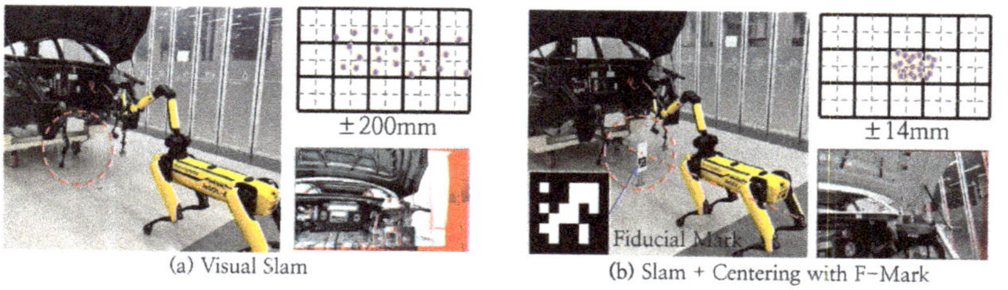

Figure 3. Improvement in repeat positioning accuracy with the centering technique using the Fiducial Mark.

4.1.2. Automatic Correction Algorithm for Image Matching Deviation Caused by Positional Precision Error

Additional hardware improvements to increase the positional accuracy would require excessive cost and time to reduce the deviation in the images acquired from the mobile robot or arm. In the industrial field, there is a trade-off between performance improvement and cost; therefore, an appropriate performance improvement method must be developed. In this study, we propose a visual software algorithm that can correct image deviation with relatively low investment cost. We used the speeded-up robust feature (SURF) algorithm to detect feature points between the first image and the repeatedly acquired images, and corrected the deviation using affine, projective, and other transformation techniques [12,13]. However, during feature point detection, there were problems with recognizing the background as a feature point instead of the inspection area, or recognizing parts incorrectly installed as the same feature point, resulting in degraded performance, as shown in Figure 4 (top). To address this issue, we limited the feature point search area to the vehicle body, as shown in Figure 4 (bottom). This prevents the recognition of the background as a feature point outside the vehicle body. Additionally, the image-matching performance is improved by masking the part area and excluding it from the search area to prevent the recognition of incorrect feature points when parts are installed incorrectly.

$$\text{argmax}\,(x, y) \in R \cap M \sum i, jH_{ij}L_{\sigma}(x+i, y+j) \quad (1)$$

$$\text{argmax}\,(x, y) \in M \sum i, jH_{ij}L_{\sigma}(x+i, y+j) \quad (2)$$

$$\text{argmax}\,(x, y) \in I \sum i, jH_{ij}L_{\sigma}(x+i, y+j) \quad (3)$$

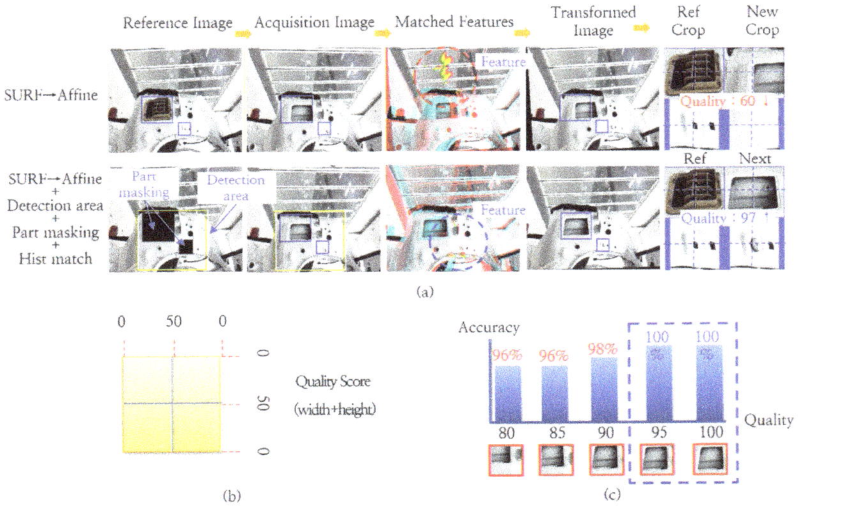

Figure 4. (**a**) Image registration and cropping using the proposed algorithm, (**b**) quality scoring of cropped images, (**c**) changes in AI accuracy according to cropped image quality.

Equation (1) is used to detect features within the intersection of the search area specified by the reference image's feature search area, R, and the image mask, M, that includes the areas outside the feature search area. H_{ij} represents the Gaussian kernel used in the Harris corner detector, and $L_{\sigma}(x+i, y+j)$ represents the result of differentiation and smoothing using the image's Rob operator with a Gaussian filter. $(x+i, y+j)$ represents the position of the kernel. Conversely, there are two ways to perform a feature search on a newly acquired image. If the image distortion is small, Equation (2) can be used, which utilizes the image mask, M, that includes the areas outside the feature search area. In this case, the area for feature detection is reduced since the same image masking area is included, resulting in an increase in the speed. However, it was observed that the accuracy of feature detection decreases with the increase in the image distortion. This is because the difference between the reference image and the masked area caused by the image distortion is severe. To improve this, I represents the entire area of the newly acquired image, enabling the feature search area to be assigned without a separate mask, as shown in Equation (3). This improves the alignment performance. However, it was also observed that the expansion of the search area increases the time required by approximately 30%. Therefore, Equation (2) must be used when the image distortion is small, and Equation (3) must be used appropriately when the distortion is significant. Consequently, only the feature points of the vehicle body that do not change before and after mounting the parts were detected, as shown in Figure 5; further, the image registration algorithms can be implemented through the image conversion methods, affine and Projective Transform, using the feature points between the two detected images [14]. The inspection image of the part was extracted from the registered image by using the ROI coordinates of the part set in the initially acquired image of SPOT, and high-quality learning data necessary for AI algorithm development could be obtained, as shown in Figure 5.

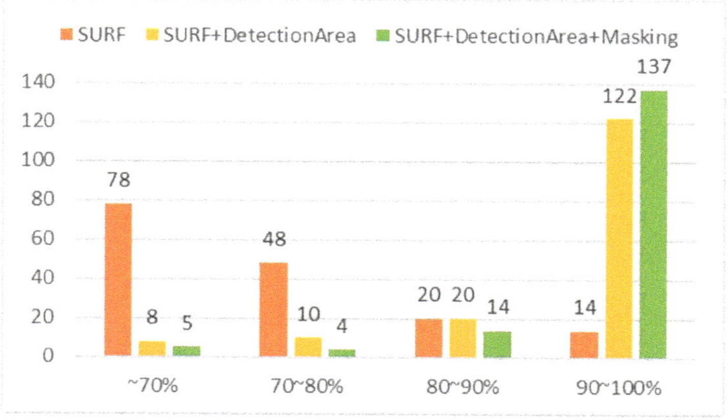

Figure 5. Evaluation of SURF algorithm performance with mixed image search area and image search exclusion area.

4.2. Image Pre-Processing Strategy for Minimizing Lighting Changes Caused by Environmental Variations

Lighting control is possible for small parts, but it is difficult to control lighting in automobile assembly processes due to the large size of the vehicle. Frequent changes in illumination occur due to the inability to control lighting. Additionally, the camera shooting angle changes due to the positional deviation of the mobile robot during image acquisition, causing variations in the gain, exposure, brightness, gamma, and other image features. Contrast changes in the image can degrade the performance of feature point detection through image comparison. Furthermore, changes in the brightness can cause excessive variations at the edges of the image, causing the performance degradation of the CNN network. To address these issues, an algorithm was applied to match the histogram of the acquired image to that of the initial reference image, thereby compensating for overexposure and brightness changes in the image [15,16].

The formula for matching each pixel value, $p_{new}(i,j)$, in the new image, I_{new}, to the histogram of the reference image, I_{ref}, as shown in Figure 6, is given as follows:

$$p_{new}(i,j) = \sum_{k=0}^{L-1} \frac{h_{ref}(k)}{h_{new}(k)} \cdot$$
$$\max(0, \min(p_{new}^{max}, k + \frac{p_{new}^{max}}{L} - p_{new}(i,j)))$$

Here, $h_{ref}(k)$ and $h_{new}(k)$ represent the histograms of I_{ref} and I_{new}, respectively. L denotes the range of pixel values and p_{new}^{max} denotes the maximum pixel value of I_{new}. This formula matches the histogram of I_{new} to that of I_{ref}, thereby improving the contrast of I_{new}. In software-based image processing after image acquisition, the original image is fixed and the range for change is set. Registration cannot be performed if there is a large difference from the original image. To solve this problem, it is more effective to acquire an image similar to the initially acquired image while changing the camera parameters during image acquisition. However, this is not suitable for automobile production plants where production cycle time is important since image acquisition time increases. Therefore, the operating time must be considered when developing a system in an industrial setting.

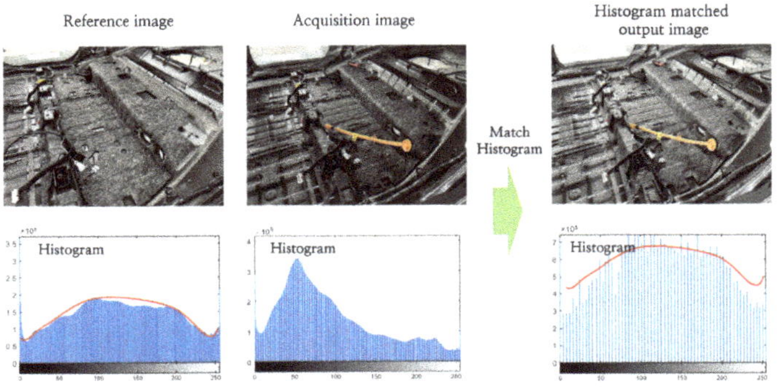

Figure 6. Image brightness correction using histogram matching algorithm.

4.3. *Development Plan for Vision-Based AI Algorithms Enabling Maintenance and Continuous Management*

In the manual assembly process of an automobile manufacturing plant, the new vehicle release cycle is fast, due to which the parts subject to inspection are frequently changed. Therefore, the cost of developing an inspection algorithm is high, and it is crucial to reduce the development period for inspection from the beginning of production. Additionally, several algorithms must be developed that can inspect a large number of parts due to mixed production, and excessive costs are incurred to maintain the performance. To effectively apply AI technology at industrial sites such as automobile manufacturing plants, maintaining appropriate development costs and reducing the development period are critical issues. In industrial settings, excessive investment costs are incurred for the re-development of algorithms when changing the inspection targets, and there are often cases where equipment is unused because it does not satisfy the required detection performance. Consequently, AI algorithms have a negative perception. In this study, we propose a development plan that can reduce the cost of developing multiple algorithms and drastically reduce the development period to effectively implement AI algorithms in industrial sites.

4.3.1. Cropping Technique to Reduce Learning Data Acquisition Time

The period for acquiring the training images must be reduced to reduce the development period of the deep learning algorithm. However, in industrial settings where product change cycles are fast, it is practically impossible to acquire learning images within a short period. Therefore, a method must be developed to acquire learning data within a short period. In this study, we aimed to maximize the reuse of learning data even when the product changes. Specifically, image pre-processing was performed to crop the inspection parts as large as possible, to exclude the highly variable vehicle structure and color from the images. This ensures that learning and inspection images can be reused even if the vehicle is changed during automobile production. Thus, even if the vehicle is changed, since the structure of parts such as the wiring, clips, connectors, and bolts assembled in the vehicle is similar, a deep learning inspection algorithm can be developed by using the learning data collected from previous vehicles. Figure 7 shows that visual inspection can be applied quickly during new car production by using the deep learning algorithm created for the previous car since the type of clip used for fixing the wiring is similar. Moreover, since the learning data are continuously accumulated and diversity is secured, the performance of the algorithm can be continuously upgraded. To verify this, when launching a new car with a similar but not the same part type, a short algorithm development test was conducted through transfer learning after securing the minimum quantity of data for the new parts that are similar. The detection performance decreases when performing deep learning with only a small number of new part images of less than 20, as shown in Figure 8. However, the results of transfer learning using the existing algorithm of similar parts exhibit higher accuracy.

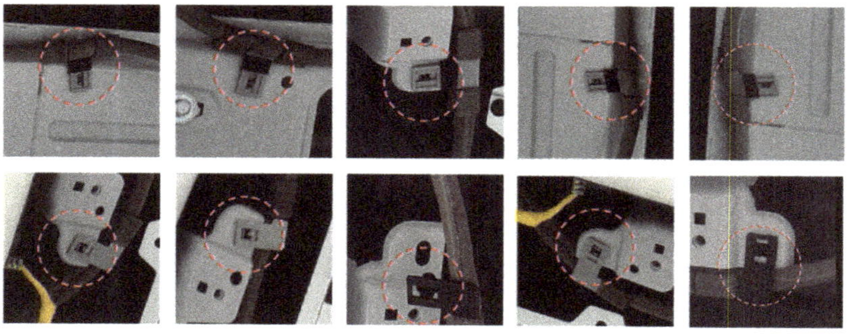

Figure 7. Inspection part unit cropping images for reuse of car assembly part types and learning images.

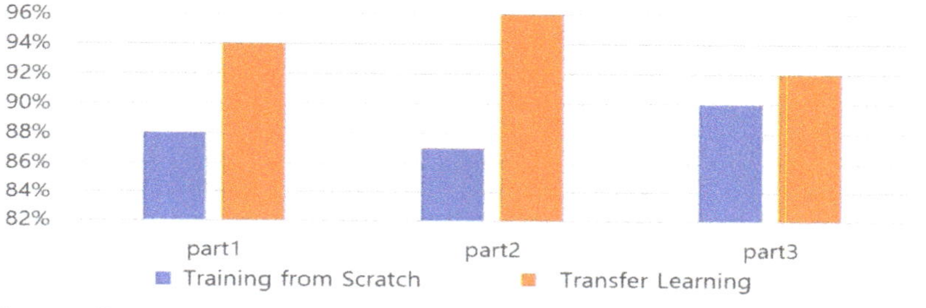

Figure 8. Comparison of results between training from scratch and transfer learning using OK (20 images) and NG (20 images) training data with the resnet101 model.

4.3.2. Minimizing the Development Period and Investment Cost of Algorithm Development Plan

Deep learning techniques are broadly classified into three categories, based on their detection performance and resource utilization: image classification, object detection, and segmentation [17]. The accuracy of the analysis increases in the order of classification, object detection, and segmentation, but the processing resources required also increase with the increase in the amount of data to be processed. Additionally, as the accuracy increases, the cost and labeling period for the training data also increase, which are crucial issues in industrial sites where there are frequent changes in the inspection items. Extensive research is being conducted on auto-labeling to address this issue [18–20]. Although this demonstrates a certain level of performance that can be achieved in industrial sites, separate confirmation is required since even one or two mislabeled data points can significantly impact the algorithm's performance. In this study, classification was applied to the data format used in manual assembly processes for inspecting the automobile parts, considering cost and data acquisition time. Since the classification technique exhibits a lower accuracy than other techniques, it must be improved. To improve the inspection accuracy, we focused on implementing image pre-processing algorithms that can obtain high-quality images and improve the algorithm performance.

4.4. Automation Technology for Maintaining the Performance of AI Algorithms

4.4.1. Automatic Image Augmentation Technology That Accounts for Deviation in Mobile Robot's Shooting Position

The image augmentation technique involves creating new data by appropriately transforming the original image during CNN deep learning training. It is effective in making the model robust when there is insufficient training data, and is important to improve the performance of deep learning algorithms [21,22]. When acquiring images using a mobile robot in an automobile assembly process, the range of image acquisition deviations caused by robot position errors can be statistically calculated. Using the calculated image deviation range value, the image augmentation range can be specified during deep learning training. Thus, the image can be augmented within the same range as the image deviation that can occur due to the positional error of the mobile robot.

Therefore, learning data with the same range of deviation as that of the inspection image can be additionally created, thereby improving the detection performance of the deep learning algorithm. Additionally, it is very effective at maintaining the algorithm performance when the error range changes due to environmental changes and robot deterioration since the algorithm can be automatically re-learned within the calculated deviation range for a certain period of time without the need for an engineer. The T-Matrix value of the image deviation information measured in the image error registration SW of Figure 1 is stored. The image acquisition deviation range for each robot position can be calculated as shown in Figure 9. It can be observed that the deviation of the error caused by the different position of the robot and the different pose of the arm for part shooting is different. Essentially, when training the deep learning algorithm using different robot position deviations for each part shooting position, an appropriate image augmentation value for each position can be used as shown in Figure 10. This method can prevent the degradation of the algorithm performance by learning with an image that is completely different from the inspection image during algorithm learning. Additionally, the error range analyzed by the image error registration software can be automatically parameterized for image augmentation without the intervention of an engineer.

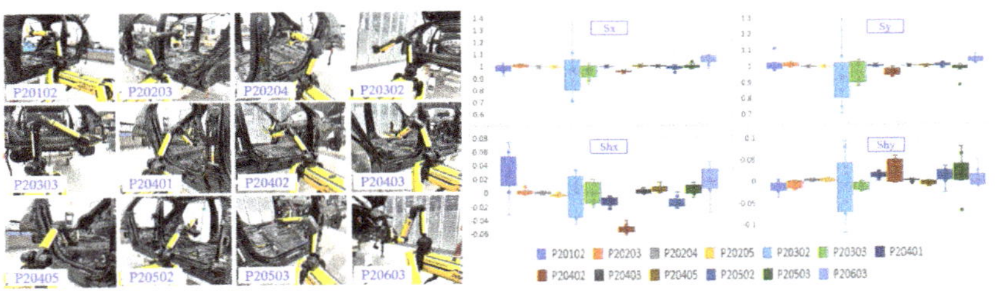

Figure 9. For each set of 100 images captured by the robot at each position, T-Matrix can be used to extract the range of augmentations.

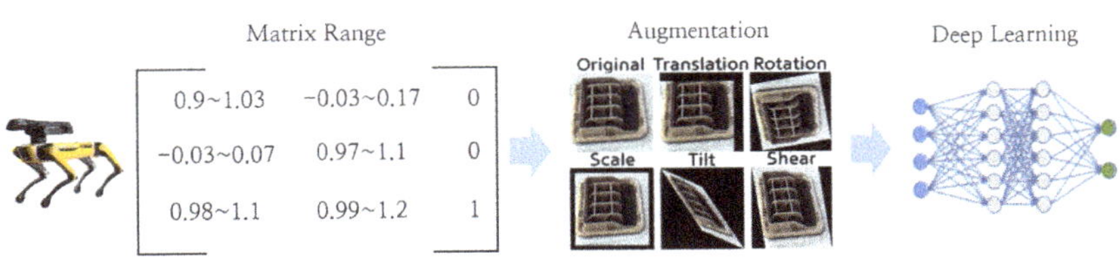

Figure 10. Using the range of image deviation caused by robot position errors as image augmentation parameters.

To verify the effectiveness of the image augmentation technique using T-Matrix values, we compared the results of learning with parameters set by engineers to the results of learning with parameters set automatically using the T-Matrix values. The results indicated equivalent algorithm performance, as shown in Figure 11. Conversely, we observed that the algorithm's performance deteriorated when the image was augmented with a difference of ±5% or more from the error range for each robot position. This implies that incorrect parameter settings by engineers during deep learning can degrade the algorithm performance, and there is a high possibility of algorithm performance deviation based on the engineer's ability. Additionally, maintenance costs can be minimized because the system can automatically learn the algorithm's performance despite environmental changes or robot deterioration, without requiring an engineer.

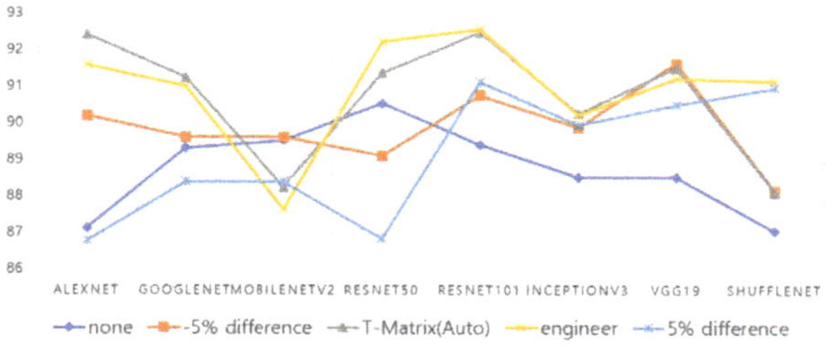

Figure 11. Comparison graph of learning accuracy for each representative network according to image augmentation error range of ±5%, T-Matrix (auto), and engineer's experience level.

4.4.2. Automatic Data Acquisition Method for Re-Learning AI Algorithms

It is difficult to implement the AI vision inspection system in the industrial field because the performance of the algorithm deteriorates due to changes in the inspection environment. When the algorithm performance degrades, it is essential to manage performance through algorithm re-learning immediately. However, it is difficult to maintain the performance of the algorithm owing to the high time and cost requirements incurred during the labeling task to transform the acquired data into learning data. Essentially, an automatic data labeling method is required to maintain algorithm performance for the implementation of AI technology in industrial sites. In this study, a classification score was used to automatically label the acquired data. To automatically secure the re-learning data, the cross entropy score value was set as high as possible in the Softmax step of the image classification process to ensure that it was OK, as shown in Figure 1. The OK data is automatically classified without operator intervention, and NG and NA, which require operator correction, can be labeled as data by clicking through a GUI that enables the operator to determine whether it is OK, NG, or NA.

4.5. Industrial Field AI Algorithm Development Plan

To improve the performance of deep learning algorithms, obtaining sufficient-quality training data is the most important aspect. However, obtaining sufficient learning data for developing vision AI algorithms within a short period that corresponds to the time of product production in industrial settings such as automobile assembly processes is extremely challenging. Since adequate data cannot be used when developing AI algorithms in industrial settings, a method must be developed to satisfy the algorithm's performance requirements using only a small amount of data that can be initially acquired. Acquiring learning data in industrial settings is very difficult, as mentioned earlier, and problems can arise due to the imbalance of the OK and NG data. The anomaly detection technique is being analyzed to solve this problem; however, it is still inadequate for implementation in industrial settings with diverse inspection images. In this study, we developed algorithms with various combinations of similar or different part images to improve the AI algorithm performance using a small amount of data, as shown in Figure 12. By selecting and using the inspection algorithm for parts with high accuracy among the developed algorithms, a high-performance detection algorithm was developed within a short period using only a small amount of learning data. To improve the algorithm detection performance, a large amount of learning data with diversity must be obtained. However, the best and easiest way to increase the diversity and accuracy of learning data is to mix between automobile parts to reduce the variance and bias, which is practically impossible to achieve under industrial settings.

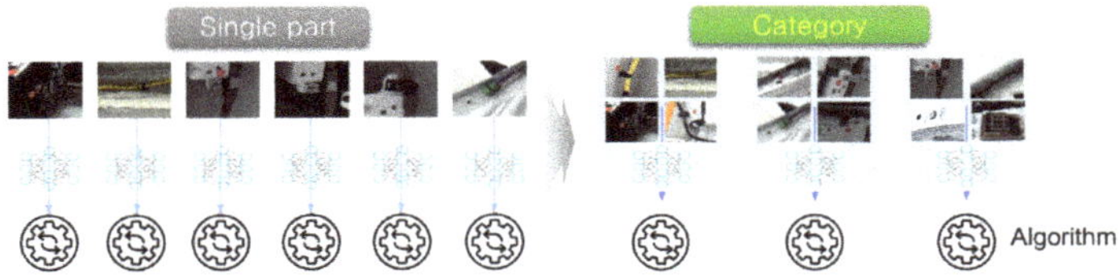

Figure 12. Create categories of similar parts, repeatedly learn with mixed categories, evaluate the accuracy of each part, and use the algorithm of the part with the highest performance.

The Boosting technique is one of the methods that can be used to solve the generalization problem of machine learning algorithms. Boosting assigns more weight to misclassified samples from previous training results on the same data to predict more accurate results [23,24]. However, Boosting has dependencies between data and network models, requiring various tuning operations to achieve the optimal performance. Conversely, the proposed learning method is very simple and intuitive, making it easy to use in industrial automation systems. Additionally, mixing similar parts of data can increase the generalization performance while reducing the variance and bias by increasing the diversity and quantity of the training data. This algorithm takes three primary inputs: the total number of car parts (N), an accuracy threshold (t) for effective model evaluation, and the number of iterations (I) to refine the models through repeated training. The output obtained is the trained models (M) for each car part that satisfy the accuracy threshold. The core process includes initializing a tracking table (T), selecting random subsets of parts for model training, comparing their accuracy against t, and iterating until the most accurate algorithms for each part are identified. After randomly mixing inspection part images as shown in Algorithm 1 and training the model, only the parts with high inspection accuracy are used in the AI algorithm. For part's with low accuracy, the algorithm is retrained using a different mix of parts until the inspection accuracy is sufficiently high to be used in the AI algorithm. Using this method, we were able to improve the performance of the AI algorithm with minimal data and resources within a short period. Using the proposed AI algorithm development method, the results of training on the same data, same network model, and same parameter setting presents higher accuracy than training for a single part, as shown in Figure 13. In particular, applying the category technique to parts with an accuracy of 0.5 or less increased the accuracy to 0.9 or higher.

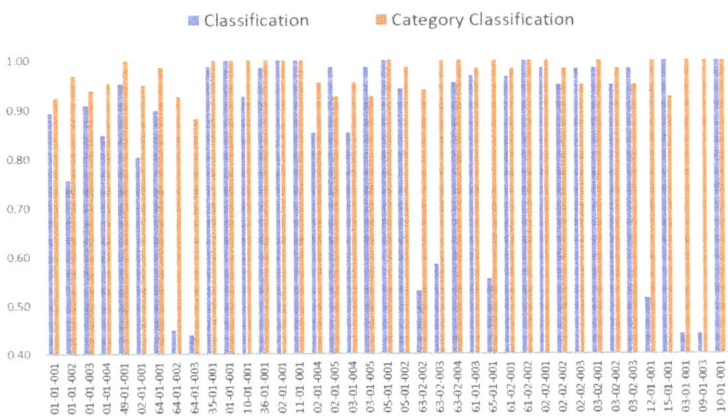

Figure 13. Performance improvement in algorithms through finding optimal AI algorithm by proposed similar part mixing.

Algorithm 1 Finding optimal algorithm for car parts

Input: Number of car parts N, accuracy threshold t, number of iterations I.
Output: Trained models M for each part.

```
Initialize an empty table T to store the part number, algorithm number, and accuracy.
for i ← 1 to I do
    for j ← 2 to N do
        Randomly select j parts to form a training dataset.
        Train a model M_j on the training dataset.
        for each part p in the j parts do
            Evaluate the model M_j on the test dataset FOR part p.
            if the accuracy of M_j for part p is above the threshold t then
                Store the part number, algorithm number, and accuracy in the table T.
            end if
        end for
    end for
end for
Group the parts by the algorithm number that achieved the highest accuracy for each part.
Save the trained model for each part and its corresponding algorithm number.
Display the table, T, with part numbers, algorithm numbers, and accuracies.
```

5. AI System Empirical Evaluation

In this study, we present the development and application of a vision AI system that can acquire images of the assembly process and inspect assembly defects using a vision AI algorithm, by utilizing the mobile robot 'SPOT'. This robot can operate in conjunction with workers on the manual assembly lines of automobile manufacturing plants. In previous studies, there was no device capable of acquiring images while operating alongside workers, and developing a solution to inspect dozens of assembly parts incurred excessive costs. Furthermore, high maintenance expenses post-system-implementation presented challenges for mass production. This paper addresses and resolves these issues. Currently, the system is being applied and operated in the prototype phase 1 process at the Singapore plant of Hyundai Motors, where it has successfully identified multiple instances of defects caused by the assembly mistakes of inexperienced workers at the initial stages of vehicle production in real-time. Additionally, this paper proposes a method for the continuous automatic collection of training data to address the lack of training data during the initial development phase. Consequently, while the average performance of the algorithm for 39 parts was initially 88%, continuous data collection and the proposed algorithm learning method have enhanced the performance to 97.4%.

6. Future Work

Several studies have been conducted on the application of AI technology in various industries. However, most of these studies approach technology development under the assumption that AI can solve everything, and this presents a major obstacle to implementing AI technology in industrial applications. To implement AI technology for improving industrial productivity, it is important to develop good deep learning networks as well as to collect training data, improve algorithm detection performance in constrained environments, maintain algorithms, and reduce the time and cost. If AI algorithms are implemented in industrial fields without such strategies, there is a high risk of failure due to real problems. Research is currently being conducted to obtain training data using 3D data because it is difficult to obtain training data at industrial sites; however, this approach is impractical [25,26]. Therefore, more realistic solutions must be developed. To enable the widespread application of AI technology in industrial fields, continuous research is required to realistically reduce the costs of developing and maintaining AI technology.

7. Conclusions

This paper analyzed the problems of industrial sites where the sustainable application of a visual AI inspection system was difficult due to frequent changes in the environmental conditions and inspection targets, and developed technologies to solve these problems. We developed image acquisition technology using the mobile robot SPOT to obtain high-quality learning data and inspection images for the real-time visual inspection of the manual assembly process of automobiles. We proposed an AI system development architecture that could be effectively applied to industrial sites. We also improved the development of AI inspection algorithms by applying technologies to reduce the learning data acquisition period, save investment costs, improve algorithm performance, and automate the algorithm maintenance. This helped in drastically reducing the existing problems. In particular, the similarity of vehicle parts was used to develop the algorithm for new B-vehicle parts by utilizing the algorithm developed for A-vehicle parts, as shown in Figure 14. If the new B-vehicle parts were identical to the A-vehicle parts, the inspection could be performed using the A-vehicle part algorithm. If the new B-vehicle parts were not identical to the A-vehicle parts but similar, the development period of the algorithm could be reduced by transferring the learning of the B-vehicle parts to the A-vehicle part algorithm. Consequently, a vision AI system with the required detection performance during the early stages of production could be applied to detect assembly defects. Furthermore, it was possible to detect many defects caused by the low skill level of workers during the initial production of new cars in automobile production factories, which could significantly improve the quality of automobile assembly. Lastly, the effective AI technology development method proposed in this study will serve as a useful guide for the implementation of AI technology in industrial sites.

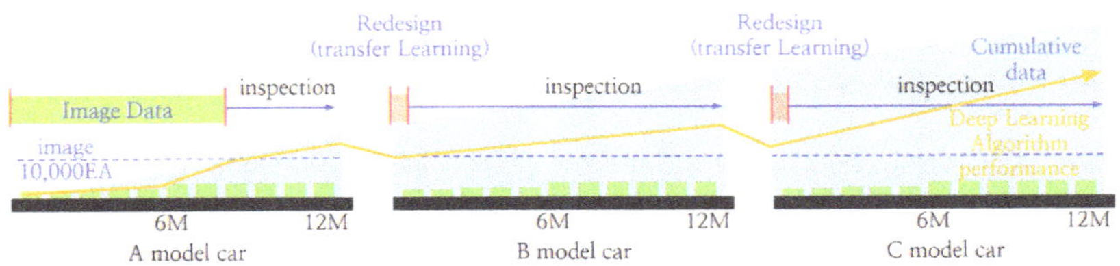

Figure 14. Algorithm development is shortened through transfer learning with same/similar part algorithms and AI algorithm improvement through continuous accumulation of automobile assembly part image data.

Author Contributions: Conceptualization, C.Y., D.K. and J.K.; Software, C.Y.; Validation, C.Y.; Formal analysis, J.K.; Investigation, J.K.; Resources, D.K.; Writing—original draft, C.Y.; Writing—review & editing, D.-S.E. All authors have read and agreed to the published version of the manuscript.

Funding: This research was supported by Hyundai Motor Company under project number 2021_CSTG_ 0174.

Institutional Review Board Statement: Not applicable.

Informed Consent Statement: Not applicable.

Data Availability Statement: The data presented in this study are available on request from the first author due to Hyundai Motor Company's internal policies and therefore are not publicly available. Data access requests can be directed to the first author.

Acknowledgments: We thank the Hyundai Motor Company for their support and resources provided for conducting this research.

Conflicts of Interest: The authors declare that they have no known competing financial interests or personal relationships that could appear to influence the work reported in this paper. However, Changmo Yang, Dongweon Kang, and JinSeok Kim are employees of Hyundai Motor Company, which provided funding and technical support for this work. The funder had no role in the design of the study; in the collection, analysis, or interpretation of data; in the writing of the manuscript; or in the decision to publish the results.

Abbreviations

The following abbreviations are used in this manuscript:

AI	Artificial Intelligence
SLAM	Simultaneous Localization and Mapping
SW	Software
ToF	Time-of-Flight
UWB	Ultra-Wideband
SURF	Speeded-Up Robust Features
ROI	Region of Interest
CNN	Convolutional Neural Network
OK	Acceptable or Correct
NG	Not Good or Incorrect
NA	Not Applicable or Not Available
AGV	Automated Guided Vehicle
T-Matrix	Transformation Matrix
F-Mark	Fiducial Mark

References

1. Yang, J.; Li, S.; Wang, Z.; Dong, H.; Wang, J.; Tang, S. Using deep learning to detect defects in manufacturing: A comprehensive survey and current challenges. *Materials* **2020**, *13*, 5755. [CrossRef] [PubMed]
2. Block, S.B.; da Silva, R.D.; Dorini, L.B.; Minetto, R. Inspection of imprint defects in stamped metal surfaces using deep learning and tracking. *IEEE Trans. Ind. Electron.* **2020**, *68*, 4498–4507. [CrossRef]
3. Mazzetto, M.; Teixeira, M.; Rodrigues, É.O.; Casanova, D. Deep learning models for visual inspection on automotive assembling line. *arXiv* **2020**, arXiv:2007.01857.
4. Hemamalini, V.; Rajarajeswari, S.; Nachiyappan, S.; Sambath, M.; Devi, T.; Singh, B.K.; Raghuvanshi, A. Food quality inspection and grading using efficient image segmentation and machine learning-based system. *J. Food Qual.* **2022**, *2022*, 5262294 . [CrossRef]
5. Lang, W.; Hu, Y.; Gong, C.; Zhang, X.; Xu, H.; Deng, J. Artificial intelligence-based technique for fault detection and diagnosis of EV motors: A review. *IEEE Trans. Transp Electrif.* **2021**, *8*, 384–406. [CrossRef]
6. Zhou, Q.; Chen, R.; Huang, B.; Liu, C.; Yu, J.; Yu, X. An automatic surface defect inspection system for automobiles using machine vision methods. *Sensors* **2019**, *19*, 644. [CrossRef] [PubMed]
7. Wang, J.; Fu, P.; Gao, R.X. Machine vision intelligence for product defect inspection based on deep learning and Hough transform. *J. Manuf. Syst.* **2019**, *51*, 52–60. [CrossRef]
8. Chalapathy, R.; Menon, A.K.; Chawla, S. Anomaly detection using one-class neural networks. *arXiv* **2018**, arXiv:1802.06360.
9. Boston Dynamics. SPOT. Available online: https://dev.bostondynamics.com/ (accessed on 21 March 2024).
10. Cheng, T.; Venugopal, M.; Teizer, J.; Vela, P. Performance evaluation of ultra wideband technology for construction resource location tracking in harsh environments. *Autom. Constr.* **2011**, *20*, 1173–1184. [CrossRef]
11. Karedal, J.; Wyne, S.; Almers, P.; Tufvesson, F.; Molisch, A.F. A measurement-based statistical model for industrial ultra-wideband channels. *IEEE Trans. Wirel. Commun.* **2007**, *6*, 3028–3037. [CrossRef]
12. Bay, H.; Ess, A.; Tuytelaars, T.; Van Gool, L. Speeded-up robust features (SURF). *Comput. Vis. Image Underst.* **2008**, *110*, 346–359. [CrossRef]
13. Bansal, M.; Kumar, M.; Kumar, M. 2D object recognition: A comparative analysis of SIFT, SURF and ORB feature descriptors. *Multimed. Tools Appl.* **2021**, *80*, 18839–18857. [CrossRef]
14. Wiki. 2D Affine Transformation Matrix. Available online: https://en.wikipedia.org/wiki/Affine_transformation (accessed on 21 March 2024).
15. Rother, C.; Minka, T.; Blake, A.; Kolmogorov, V. Cosegmentation of image pairs by histogram matching-incorporating a global constraint into mrfs. In Proceedings of the 2006 IEEE Computer Society Conference on Computer Vision and Pattern Recognition (CVPR'06), New York, NY, USA, 17–22 June 2006; Volume 1, pp. 993–1000.
16. Chen, H.M.; Varshney, P.K. Mutual information-based CT-MR brain image registration using generalized partial volume joint histogram estimation. *IEEE Trans. Med. Imaging* **2003**, *22*, 1111–1119. [CrossRef]
17. Pouyanfar, S.; Sadiq, S.; Yan, Y.; Tian, H.; Tao, Y.; Reyes, M.P.; Shyu, M.L.; Chen, S.C.; Iyengar, S.S. A survey on deep learning: Algorithms, techniques, and applications. *ACM Comput. Surv.* **2018**, *51*, 1–36. [CrossRef]
18. Fischl, B.; Salat, D.H.; Busa, E.; Albert, M.; Dieterich, M.; Haselgrove, C.; Van Der Kouwe, A.; Killiany, R.; Kennedy, D.; Klaveness, S.; et al. Whole brain segmentation: Automated labeling of neuroanatomical structures in the human brain. *Neuron* **2002**, *33*, 341–355. [CrossRef] [PubMed]
19. Gildea, D.; Jurafsky, D. Automatic labeling of semantic roles. *Comput. Linguist.* **2002**, *28*, 245–288. [CrossRef]
20. Mei, Q.; Shen, X.; Zhai, C. Automatic labeling of multinomial topic models. In Proceedings of the 13th ACM SIGKDD International Conference on Knowledge Discovery and Data Mining, San Jose, CA, USA, 12–15 August 2007; pp. 490–499.

21. Krizhevsky, A. Learning Multiple Layers of Features from Tiny Images. 2009. Available online: https://www.cs.toronto.edu/~kriz/learning-features-2009-TR.pdf (accessed on 21 March 2024).
22. Litjens, G.; Kooi, T.; Bejnordi, B.E.; Setio, A.A.A.; Ciompi, F.; Ghafoorian, M.; Van Der Laak, J.A.; Van Ginneken, B.; Sánchez, C.I. A survey on deep learning in medical image analysis. *Med. Image Anal.* **2017**, *42*, 60–88. [CrossRef] [PubMed]
23. Chen, C.; Xiong, Z.; Tian, X.; Zha, Z.J.; Wu, F. Real-world image denoising with deep boosting. *IEEE Trans. Pattern Anal. Mach. Intell.* **2019**, *42*, 3071–3087. [CrossRef] [PubMed]
24. Chen, C.; Xiong, Z.; Tian, X.; Wu, F. Deep boosting for image denoising. In Proceedings of the European Conference on Computer Vision (ECCV), Munich, Germany, 8–14 September 2018; pp. 3–18.
25. Dosovitskiy, A.; Springenberg, J.T.; Tatarchenko, M.; Brox, T. Learning to generate chairs, tables and cars with convolutional networks. *IEEE Trans. Pattern Anal. Mach. Intell.* **2016**, *39*, 692–705. [CrossRef] [PubMed]
26. Voulodimos, A.; Doulamis, N.; Doulamis, A.; Protopapadakis, E. Deep learning for computer vision: A brief review. *Comput. Intell. Neurosci.* **2018**, *2018*, 7068349. [PubMed]

applied sciences

MDPI

Article

Robot Operating Systems–You Only Look Once Version 5–Fleet Efficient Multi-Scale Attention: An Improved You Only Look Once Version 5-Lite Object Detection Algorithm Based on Efficient Multi-Scale Attention and Bounding Box Regression Combined with Robot Operating Systems

Haiyan Wang [1,2,3,*], Zhan Shi [1], Guiyuan Gao [1], Chuang Li [4], Jian Zhao [1,2,3] and Zhiwei Xu [1]

[1] College of Computer Science and Technology, Changchun University, Changchun 130022, China; zhans0125@163.com (Z.S.); gaoguiy0409@163.com (G.G.); zhaojian@ccu.edu.cn (J.Z.); xuzhiwei2000@sina.com (Z.X.)

[2] Key Laboratory of Intelligent Rehabilitation and Barrier-Free Access for the Disabled, Ministry of Education, Changchun 130022, China

[3] Jilin Provincial Key Laboratory of Human Health State Identification and Function Enhancement, Changchun 130022, China

[4] College of Computer, Jilin Normal University, Siping 136000, China; lichuang12@mails.jlu.edu.cn

* Correspondence: wanghy80@ccu.edu.cn

Abstract: This paper primarily investigates enhanced object detection techniques for indoor service mobile robots. Robot operating systems (ROS) supply rich sensor data, which boost the models' ability to generalize. However, the model's performance might be hindered by constraints in the processing power, memory capacity, and communication capabilities of robotic devices. To address these issues, this paper proposes an improved you only look once version 5 (YOLOv5)-Lite object detection algorithm based on efficient multi-scale attention and bounding box regression combined with ROS. The algorithm incorporates efficient multi-scale attention (EMA) into the traditional YOLOv5-Lite model and replaces the C3 module with a lightweight C3Ghost module to reduce computation and model size during the convolution process. To enhance bounding box localization accuracy, modified precision-defined intersection over union (MPDIoU) is employed to optimize the model, resulting in the ROS–YOLOv5–FleetEMA model. The results indicated that relative to the conventional YOLOv5-Lite model, the ROS–YOLOv5–FleetEMA model enhanced the mean average precision (mAP) by 2.7% post-training, reduced giga floating-point operations per second (GFLOPS) by 13.2%, and decreased the params by 15.1%. In light of these experimental findings, the model was incorporated into ROS, leading to the development of a ROS-based object detection platform that offers rapid and precise object detection capabilities.

Keywords: ROS; efficient multi-scale attention; C3Ghost; MPDIoU; YOLOv5-Lite

1. Introduction

Object detection is an important branch of machine vision. Its purpose is to automatically identify and locate targets of interest in images or videos. In the field of service robots, object detection technology is mainly used to identify various objects and people in the environment, so as to realize functions such as autonomous navigation, task execution, and human–computer interaction [1]. However, due to the diversity and complexity of service robot application scenarios, object detection faces many challenges, such as illumination changes, occlusion, scale changes, etc. Therefore, object detection is an indispensable function for service robots.

Due to the high accuracy and high stability of deep learning technology in image processing, many researchers have begun to use deep learning technology to solve the problem of target detection in computer vision [2]. At present, the commonly used object detection network based on deep learning can be roughly divided into the following two categories: one-stage and two-stage [3].

For two-stage object detection, Ross Girshick et al. proposed the classical region-based convolutional neural network (R-CNN) [4] algorithm. Firstly, about 2000 region proposals are obtained using selective search, then the features of region proposals are extracted by AlexNet4 [5], and then these features are regressed by multiple classifiers. Subsequently, He et al. proposed a spatial pyramid pooling network [6] (SPPNet), which extracts more feature information by performing convolution operations on the entire image to avoid the problem of computational redundancy when R-CNN extracts features for all candidate regions. Therefore, the fully connected neural network (F-CNN) [7] adds an SPPNet between the last convolutional layer and the fully connected layer to extract a fixed-length feature vector and avoid the normalization of the region proposal. Ross Girshick et al. proposed fast R-CNN [8] by referring to SPPNet, which simplifies the SPP layer to the region of interest (ROI) layer and applies singular value decomposition (SVD) to the output of the fully connected layer to accelerate the test process. Fast R-CNN combines classification with a bounding box, but fast R-CNN has the problem of excessive calculation. In this regard, Ross Girshick and others then proposed faster R-CNN [9], which uses a region proposal network (RPN) instead of a selective search algorithm to extract region proposal, which greatly improves the detection efficiency. On the basis of faster R-CNN, Lin et al. proposed the feature pyramid network (FPN) [10], which uses RPN to extract candidate regions on the feature pyramid. By fusing deep and shallow feature information, prediction is performed at different scales to enhance the semantic understanding of shallow feature maps, thereby improving the accuracy of small target detection. In order to further improve the detection speed, Dai et al. proposed a region-based fully convolutional network (R-FCN) [11], replacing the fully connected layer with a fully convolutional layer, allowing the features of each candidate region to perform convolution operations directly to obtain the confidence of each category. Although two-stage object detection has high detection accuracy, it does not perform well in real time. In this regard, target detection technology usually uses the one-stage target detection algorithm. The one-stage target detection algorithm can achieve real-time detection, and the detection accuracy can maintain the same level as the two-stage target detection algorithm [12].

The you only look once (YOLO) algorithm is an object detection algorithm that divides the trained image into a grid system. Each unit in the grid is responsible for detecting its own internal objects. The YOLO algorithm has occupied an important position in the field of target detection, with its excellent detection speed and accuracy, since it was first proposed in 2016. The YOLOv1 [13] algorithm regards the target detection problem as a regression problem, which is an end-to-end method with fast detection speed and good real-time performance. However, the detection accuracy of the algorithm is low, and it is difficult to detect when the target object is small. In order to improve this problem, Redmon et al. improved YOLOv1 by introducing batch normalization and dimension clustering to improve the detection accuracy and called the algorithm YOLOv2 [14]. On the basis of YOLOv2, Redmon et al. further improved it through a series of improvements, such as using a residual network to improve the network structure to achieve multi-scale output, thereby improving the accuracy of detection, and named the improved algorithm YOLOv3 [15]. Since the accuracy rate has been greatly improved after using the YOLOv3 algorithm, it has become one of the most used algorithms. Bochkovskiy et al. proposed the YOLOv4 algorithm [16] for some shortcomings in the YOLOv3 algorithm and made a series of improvements. The algorithm uses the cross-stage partial darknet-53 (CSP-Darknet53) [17] structure to optimize the network structure and uses the data enhancement method in the training phase to further improve the training speed and accuracy. In 2020, Ultralytics developed an open-source version. The backbone network of the YOLOv5 algorithm [18] takes into account both the detection efficiency and image recognition effect. The volume of the algorithm model can be adjusted, and the recognition result is more accurate than other detection methods, but the calculation amount of the model is large, and the structure is redundant. In response to these challenges, this paper proposes an improved YOLOv5-Lite target detection algorithm that combines multi-scale attention and bounding box regression, aiming to further improve the detection performance while maintaining the lightweight characteristics of the algorithm. As a widely used robot software platform, ROS1 provides a wealth of tools and libraries to support the development and integration of algorithms. The improved YOLOv5-Lite algorithm is integrated with ROS, which can not only realize the rapid deployment of the algorithm but also facilitate the interaction with other robot perception and decision-making modules through the modular characteristics of ROS. Next, the design and implementation of the improved YOLOv5-Lite algorithm will be introduced in detail, as well as the experimental process and result analysis in the ROS environment.

2. Related Work

As a one-stage object detection algorithm, the YOLO series algorithm has high detection accuracy and achieves a good balance between accuracy and recognition, which is suitable for object detection in complex natural environments [19]. As the latest lightweight version of this series, YOLOv5-Lite is designed for computing resource-constrained environments. It provides acceptable accuracy while maintaining high detection speed. Although YOLOv5-Lite performs well in some application scenarios, there is still room for improvement in specific robot vision tasks, for example, the detection accuracy of small targets in a dynamic environment, or the robustness under different lighting conditions.

2.1. YOLOv5-Lite Network Model

The YOLOv5-Lite model adopts a lightweight design to reduce computational complexity and improve operating efficiency while maintaining high detection accuracy. This structural optimization makes the algorithm more suitable for running on resource-constrained devices. The network structure is shown in Figure 1.

The structure can be roughly divided into the backbone network, neck network, and detection head network. The algorithm removes the focus structure layer, reduces the volume of the model, and makes the model lighter; at the same time, four slice operations are removed, which reduces the occupation of the computer chip cache and reduces the processing burden of the computer. Compared with the YOLOv5 algorithm, the YOLOv5-Lite algorithm can avoid repeated use of the C3 layer module [20]. The C3 layer module will occupy a lot of running space on the computer, thus reducing the processing speed. In this way, the accuracy of the YOLOv5-Lite algorithm model can be controlled within a reliable range, making it easier to deploy. At the beginning of the backbone, YOLOv5-Lite uses the Conv_Batch_Norm_ReLu structure [21] to replace the traditional focus structure.

The deep stacking module of ShuffleNet V2 [22] divides the input feature channels into two parts directly through the channel splitting function. The left side does not participate in convolution and is constant, which plays the role of residual edge. After feature fusion, a channel shuffle is performed, and the left and right features can be effectively communicated. Because the down-sampling module changes the size of the feature map, a deep separable convolution is also added to the original residual edge on the left side, and the number of feature channels is changed so that the two sides after convolution can be fused. The two modules finally fuse and communicate the features after grouping through channel shuffle. The basic unit of ShuffleNet V2 is shown in Figure 2.

The ShuffleNet V2 model has a good trade-off between the speed and accuracy of image recognition. At the expense of certain prediction accuracy, a faster inference speed and smaller model parameters are obtained. YOLOv5-Lite uses a large number of Shuffle_Block operations in its backbone, which can reduce memory access, reduce the number of convolution operations, and meet lightweight design requirements.

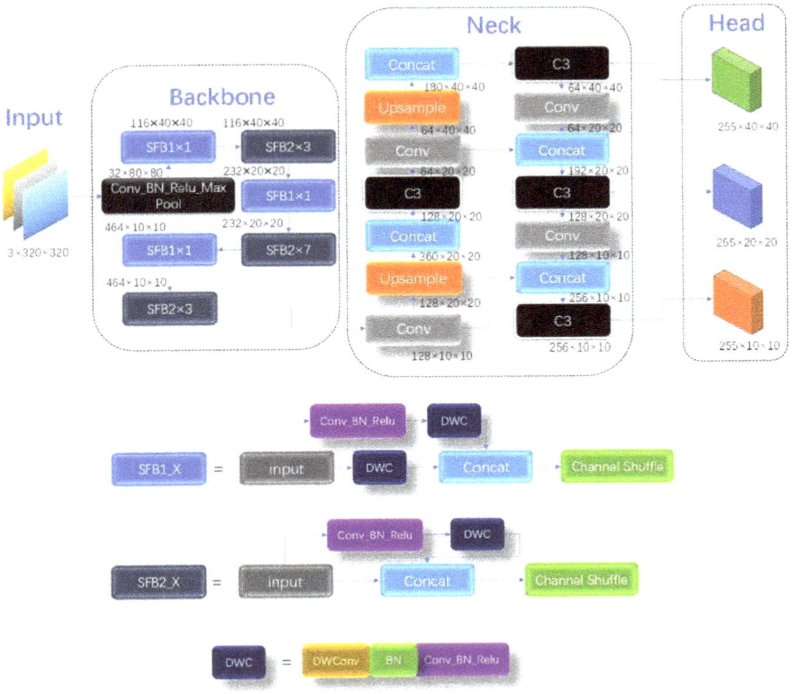

Figure 1. YOLOv5-Lite network structure.

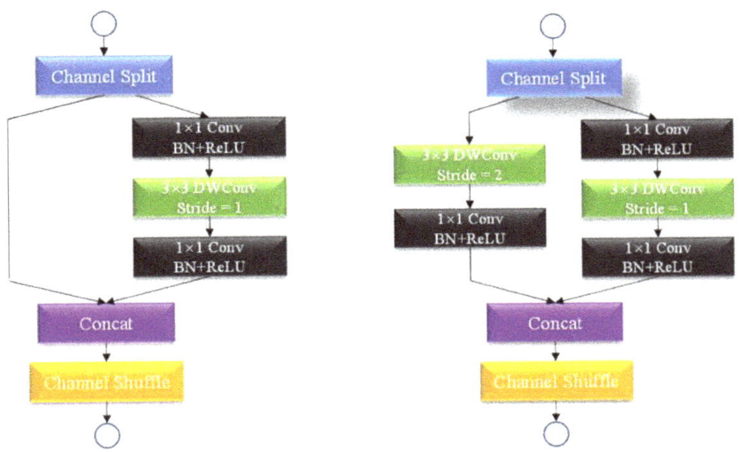

(a) (b)

Figure 2. Basic units of ShuffleNet V2. (**a**) Deep stacking module Stage 1; (**b**) deep stacking module Stage 2.

2.2. *Efficient Multi-Scale Attention*

Scientists' research on human vision shows that the human brain only selectively extracts the visual information of its own region of interest while ignoring the visual information of other regions. For example, when reading, humans will only focus on some key words and ignore some non-key words. In recent years, deep learning scholars have used the method of human brain processing vision to apply this attention mechanism to deep learning models. The experimental results show that the performance of the model can be improved to a certain extent.

EMA [23] is an efficient multi-scale attention mechanism, which reshapes some channels into batch dimensions, thereby avoiding the situation of channel dimension reduction so as to retain the information of each channel and reduce the computational cost. EMA not only adjusts the channel weight of parallel sub-networks using global information coding but also fuses the output features of two parallel sub-networks through cross-latitude interaction. The overall structure of EMA is shown in Figure 3.

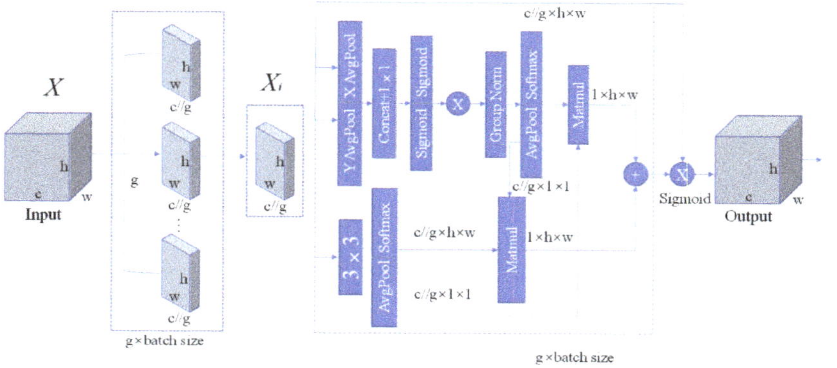

Figure 3. Efficient multi-scale attention.

In the figure, "c" denotes the number of channels in the input feature map, "h" and "w" represent the height and width of the feature map respectively, "g" denotes the number of groups, "X Avg Pool" denotes the 1D horizontal global pool, and "Y Avg Pool" denotes the 1D vertical global pool. The expression "c//g×h×w" calculates the total number of elements in the entire feature map by first dividing the channel count (c) by the group count (g) to determine the channels per group and then multiplying this value with the height (h) and width (w) of the feature map.

For the input features, EMA divides them into g sub-features according to the number of channels to learn different semantics. Without losing generality, it is assumed that the learned weight descriptor will be used to enhance the feature representation of the region of interest in each sub-feature.

In deep learning, convolution kernels of different sizes can capture features at different scales. Convolutional kernels sized 1 × 1 are typically used to capture fine-grained detail information, while 3 × 3 convolutional kernels can capture a wider range of contextual information. By combining these two sizes of convolution kernels, EMA can simultaneously obtain local and slightly global features, thereby enhancing the expressive power of the features.

EMA extracts the weight descriptor of the grouping feature map through two parallel paths on the 1 × 1 branch and one on the 3 × 3 branch. In the 1 × 1 branch, two 1D global average pooling operations are used to encode the channel along two spatial directions, and the two coding features are connected so that it does not reduce the dimension on the 1 × 1 branch. Then, the output after 1 × 1 convolution is re-decomposed into two vectors, and two Sigmoid nonlinear functions are used to fit the 2D binary distribution on the linear convolution. Finally, the cross-channel interaction is realized by multiplying the channel attention. In the 3 × 3 branch, a 3 × 3 convolution is used to capture the multi-scale feature representation.

The 2D global average pooling is used to encode the global spatial information in the outputs of 1 × 1 branches and 3 × 3 branches. The output will be converted into the corresponding dimension shape. Finally, the nonlinear function Softmax is added to fit the linear transformation. The output of the same size of the two branches is connected and converted into the R1 × H × W format. The matrix dot product operation is used to multiply the results of the above parallel processing to obtain a spatial attention map, which can collect spatial information at different scales. The final output of EMA is the same size as the input X, which is convenient to be directly added to the YOLOv5-Lite network.

2.3. MPDIoU Loss Function

Bounding box regression (BBR) [24] has an important influence on the accurate positioning and recognition of the model and is the key link to achieving efficient and accurate object detection. At present, most of the existing BBR loss functions can be divided into the following two categories: loss function based on ln norm and loss function based on intersection over union (IoU). The traditional bounding box regression loss function has the same aspect ratio in the prediction box and the actual annotation box, so it cannot be optimized. The MPDIoU loss function combines the concept of minimum point distance and improves the regression efficiency and accuracy by minimizing the distance between the upper left and lower right points between the prediction box and the real box. This process can be described as follows:

$$d_1^2 = \left(x_1^B - x_1^A\right)^2 + \left(y_1^B - y_1^A\right)^2 \tag{1}$$

$$d_2^2 = \left(x_2^B - x_2^A\right)^2 + \left(y_2^B - y_2^A\right)^2 \tag{2}$$

$$MPDIou = \frac{A \bigcap B}{A \bigcup B} - \frac{d_1^2}{w^2 + h^2} - \frac{d_2^2}{w^2 + h^2} \tag{3}$$

The parameters A and B denote two arbitrary convex images, w is the width, and h is the height; (x_1^A, y_1^A) and (x_2^A, y_2^A) represent the coordinates of the upper left corner and the lower right corner of A, respectively; (x_1^B, y_1^B) and (x_2^B, y_2^B) represent the coordinates of the upper left corner and the lower right corner of B, respectively; d_1^2 is the square of Euclidean distance between the upper left corner of A and B; d_2^2 is the square of Euclidean distance between the lower right corner points of A and B; and MPDIou is the intersection and union ratio (IoU) of A and B minus the normalized minimum point distance.

2.4. C3Ghost Module

GhostNet is a new lightweight deep neural network architecture proposed by Huawei Noah's Ark Laboratory [25]. In general, a large number of redundant feature maps generated by convolution have little complementary effect on the main feature maps in the actual detection task, which is not helpful for the network to improve detection accuracy. However, generating these redundant feature maps consumes a lot of computing power. Therefore, GhostNet constructs the ghost module and uses it to generate redundant feature maps faster and more efficiently. The GhostNet lightweight network can greatly reduce the amount of calculation and parameters of the network while maintaining the size and channel size of the original convolution output feature map. The implementation principle is to divide the traditional convolution into two steps, which are ordinary convolution and cheap linear calculation. Firstly, a part of the feature map is generated by using fewer convolution kernels, then the channel convolution is performed on this part of the feature map to generate more feature maps, and finally, the two sets of feature maps are spliced to generate the GhostNet feature map. The traditional convolution and GhostNet convolution processes are shown in Figure 4.

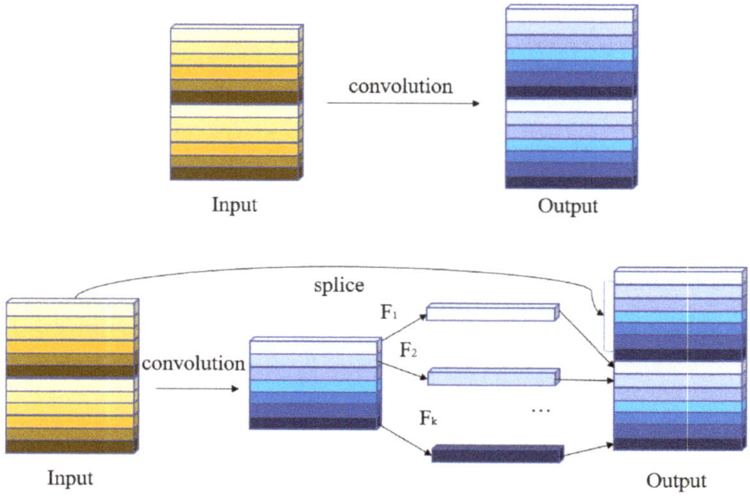

Figure 4. Traditional convolution and GhostNet convolution processes.

The head part of YOLOv5-Lite adopts multiple C3 structures, which have a large number of parameters and a slow detection speed. Therefore, this study replaces the new C3Ghost module with the C3 module to achieve a lightweight effect. The specific structure is shown in Figure 5.

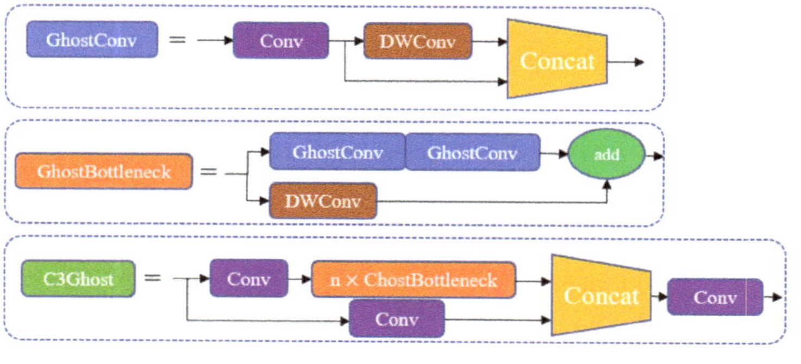

Figure 5. Ghost module.

The GhostBottleneck module is an innovative network structure component, and its design inspiration comes from the ghost module. The module is mainly composed of two GhostConv modules and a residual block. Among them, the first GhostConv module acts as an extension layer, and its core function is to increase the number of channels of the input feature map. This step is crucial because it provides a richer feature representation for subsequent deep feature extraction and information fusion. The second GhostConv module undertakes the task of dimensionality reduction, which aims to reduce the number of channels of the output feature map. This not only helps to reduce the computational complexity but also ensures that the output feature map matches the other structures in the network (such as the diameter structure) in the number of channels so as to achieve more efficient information transmission and processing. Between the two GhostConv modules, a residual edge with deep convolution processing is also embedded. This design enables the features to be effectively fused with the features of the residual edge after expansion and dimensionality reduction, thereby enhancing the expression ability of the model. In addition, the main purpose of introducing depthwise convolution (DWConv) is to further reduce the number of parameters of the model, thereby reducing the computational burden and improving the practicability of the model. The C3Ghost module is an improvement based on the C3 block. It replaces the traditional residual component Resunit with a reusable GhostBottleneck module. This replacement not only reduces a large number of convolution operations in the traditional structure but also significantly compresses the size of the model and reduces the computational complexity of the model. In this way, the C3Ghost module achieves a lightweight model while maintaining its performance, making it more suitable for deployment and operation in resource-constrained environments. This design not only improves the efficiency of the model but also enhances its adaptability and flexibility in practical applications.

3. Experimental Test and Result Analysis

The background of the robot system and the related technologies to realize ROS robot object detection are described in the previous section. Combined with the above technology, EMA is inserted into the YOLOv5-Lite model, and a lightweight C3Ghost module is designed to replace the C3 module in the traditional network to compress the calculation amount and model size of the convolution process. In order to further improve the positioning accuracy of the bounding box, the MPDIoU loss function is used to optimize the ROS–YOLOv5–FleetEMA algorithm. This chapter will introduce the basic service platform of object detection built during the experiment and train the model of the algorithm proposed in this paper. The contrast experiment and ablation experiment are designed. Finally, the object detection technology is integrated and deployed to the robot equipment for testing. The experimental process and experimental results are as follows.

3.1. Hardware Equipment

In this paper, an Ackerman differential car integrated with ROS is selected as the experimental equipment. The robot integrates a variety of sensors and computing equipment. It is equipped with laser radar for environmental perception, a camera for visual information capture, an inertial measurement unit (IMU) for attitude and motion information, a motor with an encoder for the precise control of motion, and embedded computing hardware for data processing and algorithm execution. Servo motors and stepper motors are used to precisely control the motion of robots. IMU can provide data on robot acceleration and angular velocity, which is crucial for robot positioning and navigation. Laser lidar is commonly used to detect static and dynamic obstacles. The camera is the main visual sensor for object detection, which is used to capture two-dimensional images of the scene. Through the image processing and computer vision algorithms of Raspberry Pi 4B, objects in images can be recognized and classified. The detailed layout and configuration of the hardware structure are shown in Figure 6.

3.2. Experimental Equipment

This paper uses the ROS melodic version, and the corresponding Ubuntu version is 18.04, which is installed on a virtual machine.

3.2.1. SSH Remote Connection

When we are debugging the car, we usually need to run the command line on the ROS host. However, if the display, keyboard, mouse, and other input devices are directly connected to the car to operate, when the car is in the process of movement, this method would be very inconvenient and may even affect the safety and efficiency of the operation. In order to avoid this situation and to ensure that flexible debugging and control are still possible when the car is moving, we adopted the method of remote control to realize the control of the car.

Usually, we use secure shell (SSH) login for remote control. SSH is a widely used network protocol that provides security for remote login sessions and other network services. Through SSH, we can safely execute commands on the remote host on the local computer, just like operating directly next to the car.

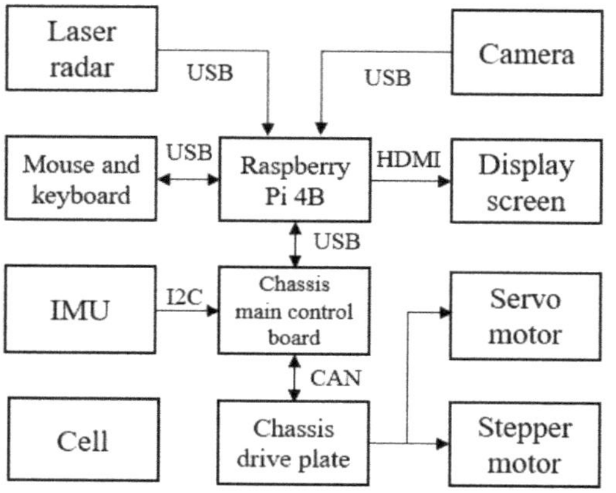

Figure 6. Hardware structure of Ackerman differential car.

3.2.2. Deep Learning Environment

First, install miniforge3 for the car; after installation, enter the following command to create a virtual environment:

- conda create -n yolo python=3.8
- conda activate yolo
- conda install pytorch torchvision torchaudio cpuonly-c pytorch

3.3. ROS-Based Object Detection Service Platform

In order to simplify the complex compilation and parameter modification process in the use of ROS, this part designs an object detection service platform based on ROS. The platform combines Qt and ROS technology; through the intuitive graphical user interface (GUI) [26], the use of buttons, input boxes, and other controls to achieve a key operation greatly improves the user experience and operational efficiency. For example, users can easily complete the SSH login device, mount the device file, open the object detection function, and conduct other operations by clicking the button, making the debugging process more convenient and clear.

After ensuring that the host and the car are in the same network environment, start the software; first, click the SSH button to remotely connect the device. Since the SSH password-free login has been configured, this step does not require additional password input, thereby simplifying the connection process. Next, in order to view and modify the car's source files in the virtual machine, you need to use a network file system (NFS) mount to mount the device's files to the virtual machine. Just click the NFS button, and you can automatically complete the file mount operation. After the file is mounted, it first needs to enter the object detection deep learning environment and then start the object detection function. To this end, this section developed a powerful target detector. Firstly, a node handle is created, and it is used to create image transmission objects and subscribe topics. When a new message is received, the callback function is called to convert the ROS image information into Opencv format, and then the function is used to convert the Opencv format object into Qt's QImage object of different depths according to the depth and channel number of cv.Mat so as to realize the visual display of the image. The workflow is shown in Figure 7.

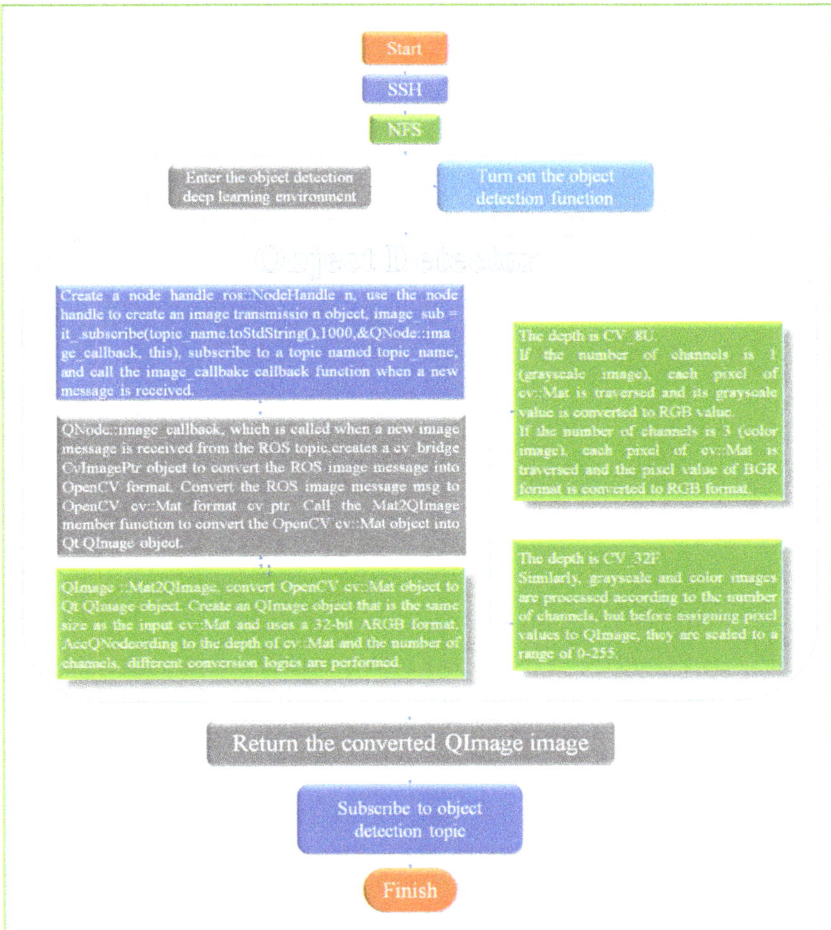

Figure 7. Workflow of object detection function.

3.4. Experimental Test

Based on the abovementioned platform technology, this part will carry out model training on the algorithm proposed in this paper, design comparative experiments and ablation experiments, and finally, integrate the object detection technology into the robot equipment for testing. The experimental process and experimental results are as follows.

3.4.1. Model Training

The computer configuration used for model training is shown in Table 1.

In this paper, a series of training parameters and strategies are used to ensure that the model can learn efficiently and stably. The weight used for training is v5lite-s.pt. The specific file and label path are input through the yaml file, in which the Batch_size is set to 16, the iteration period epochs are set to 150, the confidence threshold is 0.45, and the iou threshold is 0.65. The warmup learning rate method is used for training. When the model is in the initial epochs, a smaller learning rate value will be selected to increase the stability of the model in the initial training stage. After stabilization, the training will be continued with a preset cyclic learning rate to improve the convergence speed. The hyperparameter settings in the training process are shown in Table 2.

Table 1. Experimental environment.

Environment Configuration	Name	Related Configuration
Hardware environment	CPU	Intel(R) Core (TM) i7-7700HQ CPU@2.80GHz
	Running memory	8 G
	GPU	NVIDIA GeForce GTX 1050Ti
Software environment	Operating system	Windows10
	Python	3.8
	Deep learning framework	Pytorch
	CUDA	11.3

Table 2. Hyperparameter settings.

Parameter	Parameter Description	Value
lr0	Initial learning rate	0.001
lrf	cyclical learning rates	0.2
weight_decay	Weight attenuation parameter is used to prevent model over-fitting	0.0005
warmup_epochs	Warmup learning rounds	3.0
momentum	Warmup learning momentum	0.8
warmup_bias_lr	Warmup learning rate	0.1
IoU loss coefficient	It is used to measure the overlap between the predicted bounding box and the real bounding box	0.05
cls loss coefficient	It is used to measure the prediction accuracy of the model for the target category	0.5
cls BECLoss	Positive sample weight	1.0

3.4.2. Dataset

The pattern analysis, statistical modeling, and computational learning visual object classes challenge (PASCAL VOC) represents a challenge in the field of international computer vision. This paper selects PASCAL VOC 2007. PASCAL VOC 2007, as one of the early datasets, holds an important historical position in the field of computer vision. PASCAL VOC 2007 covers over 12,000 labeled objects and is one of the commonly used standard test datasets for the YOLO algorithm. Due to its high-quality annotation information and various common target categories, it is more consistent and accurate in category definition and annotation, including more occlusions or more complex scene layouts, making it suitable for robot object detection and fully testing the performance of the model.

3.4.3. Comparative Experiment

The specific indicators for evaluating the performance of the algorithm in this paper include precision (P), recall (R), and mAP, where mAP @ 0.5 represents the average precision mean when the intersection over the Unio IOU (IOU) threshold is 50% and mAP0.5–0.95 represents the average precision mean of the IOU threshold in the range from 50% to 95%. This process can be described as follows:

$$P = \frac{TP}{TP + FP} \tag{4}$$

$$R = \frac{TP}{TP + FN} \tag{5}$$

$$\text{mAP} = \frac{\sum_{k=1}^{n} PR}{N} \tag{6}$$

In the formula, TP represents the number of correct positive samples, FP represents the number of wrong positive samples, FN represents the number of wrong negative samples, and N represents the number of types in the sample.

Params are an important indicator for measuring model complexity. The more parameters a model has, the more computing resources and data it requires for training and inference. GFLOPS stands for the computational efficiency and speed of a model. It denotes the number of floating-point operations needed to run a network model once. It measures the number of floating-point operations a model performs during a forward propagation.

In order to verify the effectiveness of the improvement of the attention mechanism, the added attention mechanism is replaced, including squeeze-and-excitation (SE) [27], the convolutional block attention module (CBAM) [28], efficient channel attention (ECA) [29], coordinate attention (CA) [30], and EMA. The effects of different attention mechanisms on the model detection effect are compared and analyzed. The "-" indicates that the attention mechanism is not applied to the model. The same parameters are used in the training process, and experiments are performed on the VOC dataset. The results are shown in Table 3.

Table 3. Comparison of attention improvement effects.

Model	mAP @ 0.5	mAP @ 0.5–0.95	Precision	Recall
-	0.765	0.515	81.2	66.8
SE	0.761	0.523	78.4	68.1
CBAM	0.763	0.505	83.6	60.5
ECA	0.768	0.525	83.2	63.9
CA	0.769	0.521	73.7	66.3
EMA	0.776	0.534	85.6	69.1

SE is a classic channel attention mechanism, which strengthens the importance of feature channels by compressing and stimulating processes. As another form of channel attention, ECA enhances feature representation by effectively capturing cross-channel correlations but ignores spatial location information. CA integrates location information into channel attention and processes features in different spatial directions through two feature coding steps, thereby generating weights that fuse channel and spatial information. CBAM combines the advantages of channel and spatial attention mechanisms and models the channel and spatial weights independently, which not only strengthens the relationship between channels but also considers the spatial interaction and realizes the comprehensive optimization of features. The experimental results reveal the specific effects of different attention mechanisms on model performance. The model with SE and CBAM attention mechanisms suffered a 0.4% and 0.2% decrease in detection accuracy, respectively, indicating that the two mechanisms did not effectively improve performance on the current dataset. In contrast, when the model combines the CA, ECA, and EMA attention mechanisms, the detection accuracy is improved by 0.4%, 0.3%, and 1.1%, respectively. For the latter two attention mechanisms, the detection accuracy is significantly improved. On the whole, the introduction of the EMA attention mechanism not only accelerates the detection speed but also effectively improves the detection accuracy of the model, which makes it more advantageous in practical applications.

In order to verify the performance of the ROS–YOLOv5–FleetEMA model proposed in this paper, the model is compared with the traditional YOLOv5-Lite model based on deep learning. In the case of using the same dataset and experimental environment, the average accuracy improvement effect is shown in Table 4.

Through the analysis of the results, the mAP @ 0.5 of the ROS–YOLOv5–FleetEMA model proposed in this paper is 2.7% higher than that of the traditional YOLOv5-Lite model, and in a wider accuracy range mAP @ 0.5–0.95, the ROS–YOLOv5–FleetEMA model proposed in this paper is 4.3% higher than the traditional YOLOv5-Lite model.

In order to evaluate the lightweight improvement effect of the ROS–YOLOv5–FleetEMA model more comprehensively, this paper introduces the traditional YOLOv5 s model as the comparison benchmark. The experimental results are shown in Table 5.

Table 4. mAP improvement effect comparison.

Model	mAP @ 0.5	mAP @ 0.5–0.95
YOLOv5-Lite	0.765	0.515
ROS-YOLOv5-FleetEMA	0.792	0.558

Table 5. Lightweight improvement effect comparison.

Model	GFLOPS	Param
YOLOv5s	15.9	7,064,065
YOLOv5-Lite	3.8	1,566,561
ROS–YOLOv5–FleetEMA	3.3	1,332,471

The results show that the ROS–YOLOv5–FleetEMA model proposed in this paper has achieved significant optimization in the two key indicators of GFLOPS and Param. Compared with the traditional YOLOv5s model, the GFLOPs of the ROS–YOLOv5–FleetEMA model are reduced by 79.3%, and the parameter amount is reduced by 81.1%. This optimization not only reduces the consumption of computing resources but also makes the model more suitable for deployment on resource-constrained devices. At the same time, compared with the YOLOv5-Lite model, the GFLOPs of the ROS–YOLOv5–FleetEMA model are reduced by 13.2%, and the amount of parameters is reduced by 15.1%.

By comparing the experimental results, it is verified that the ROS–YOLOv5–FleetEMA model shows significant advantages in computational efficiency and resource consumption while maintaining high detection accuracy, which proves its practicability and effectiveness in a resource-constrained environment.

3.4.4. Ablation Experiment

In order to further verify the effectiveness of the improved method ROS–YOLOv5–FleetEMA model proposed in this paper, the following ablation experiments are designed: Conduct ablation experiments to explore the effectiveness of improvement methods on the model. Combine EMA, C3Ghost, and MPDIoU with the traditional YOLOv5-Lite model in different ways. By comparing the performance of models with different configurations, ablation experiments can help us understand how each component affects the overall performance of the model, including detection accuracy, computational efficiency, and resource consumption. The ablation experiment systematically removes or replaces various components in the model, observes the impact of these changes on model performance, provides an empirical basis for model design decisions, and ensures the practicality and effectiveness of the proposed ROS–YOLOv5–FleetEMA model in resource-constrained environments. In the experimental design, "−" indicates that an improvement has not been applied to the model, while "+" indicates that the improvement has been integrated. In this way, the specific impact of each combination on the performance of the model can be clearly demonstrated. The specific results are shown in Table 6 and Figure 8.

Through experimental analysis, the EMA attention module is introduced into the traditional YOLOv5-Lite model, and the mAP @ 0.5 is significantly improved, while the number of model parameters does not increase much. In addition, the traditional CIoU loss function is replaced by the MPDIoU loss function, which further optimizes the performance of the model in terms of bounding box positioning accuracy. The MPDIoU loss function makes the model more accurate in predicting the bounding box by considering the center point and diagonal distance of the bounding box, and the predicted bounding box has a higher degree of coincidence with the real bounding box. The experimental results show that mAP @ 0.5 is increased by 0.4%, which indicates that the MPDIoU loss function can make the regression of the model to the bounding box more stable, and the prediction accuracy is higher. After the introduction of the C3Chost module, the parameters of the model and the GFLOPS are significantly reduced while maintaining a high detection accuracy. The C3Chost module reduces the consumption of computing resources by optimizing the feature extraction process without affecting the detection effect. Finally, all these improved methods are applied to the YOLOv5-Lite model; not only has mAP @ 0.5 been significantly improved but the number of parameters of the model has been reduced by 15.1%. This shows that these optimization strategies can significantly reduce the computational complexity and resource consumption of the model without sacrificing the detection accuracy, making the model more suitable for deployment on resource-constrained devices, such as mobile devices and embedded systems.

Table 6. Ablation experimental results.

Method	EMA	C3Ghost	MPDIoU	mAP @ 0.5	mAP @ 0.5–0.95	GFLOPs	Param
YOLOv5-lite	-	-	-	0.768	0.515	3.8	1,566,561
YOLOv5-lite + EMA	+	-	-	0.776	0.534	3.8	1,566,575
YOLOv5-lite + C3Ghost	-	+	-	0.769	0.519	3.3	1,328,617
YOLOv5-lite + MPDIoU	-	-	+	0.772	0.522	3.8	1,566,561
YOLOv5-lite + EMA + C3Ghost	+	+	-	0.783	0.539	3.3	1,328,617
YOLOv5-lite + EMA + MPDIoU	+	-	+	0.778	0.536	3.8	1,566,575
YOLOv5-lite + C3Ghost + MPDIoU	-	+	+	0.771	0.533	3.3	1,328,617
ROS–YOLOv5–FleetEMA	+	+	+	0.792	0.558	3.3	1,332,471

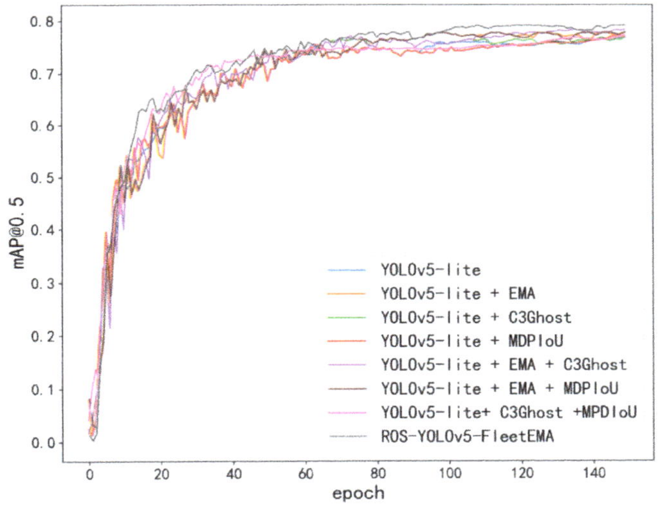

Figure 8. Ablation experimental results.

3.4.5. Integrated Experiment

The Jilin Provincial Key Laboratory of Human Health Status Identification and Function Enhancement was selected as the experimental site.

In order to realize the object detection function, this paper deploys deep learning object detection technology to the robot equipment. For this reason, this paper develops an object detection function package based on ROS–YOLOv5–FleetEMA, enters the src directory in the working space catkin_ws, and opens the terminal; input conda activate yolo, enter the virtual environment, enter the function package directory, enter sudo pip install -r requirements.txt, and install the object detection-related dependency library.

After the installation is completed, enter the roslaunch yolov5_ros yolo.launch command and start the usb_cam and the object detection function based on ROS–YOLOv5–FleetEMA at the same time. The usb_cam is a package used for interacting with the USB camera. This package allows users to subscribe to camera image topics and publish them to ROS, allowing them to use USB cameras in ROS. By subscribing to the image topic published by usb_cam, we employ cv-bridge to transform ROS image messages into the OpenCV image format. Within the callback function, we execute YOLOv5 object detection on the transformed image, subsequently convert the processed image back into ROS image messages, and publish them to a new YOLOv5 topic.

At this point, open the ROS-based object detection service platform, set the IP address of the car, and then connect the device; by using the QT button to subscribe to newly established YOLOv5 topics with just one click, the results will be displayed on the ROS-based object detection platform, as shown in Figure 9.

Figure 9. ROS-based object detection platform.

Through experimental analysis, the application effect of the ROS–YOLOv5–FleetEMA model proposed in this paper in the ROS robot system is verified. The model not only performs well in a resource-constrained environment but also integrates with a ROS-based object detection platform to achieve efficient and fast object detection. Specifically, the system can accurately identify and track multiple targets, such as pedestrians, monitors, etc. When the car maintains a speed of 0.5 to 1.5 m per second, it takes 34.4 milliseconds to identify the object, up to 30 FPS, ensuring the fluency of the detection process. This optimization not only improves the robot's perception ability in complex environments but also provides strong support for further decision-making and execution.

4. Conclusions

This paper comprehensively introduces the development process of ROS and makes an in-depth analysis of object detection technology. Along with the hardware equipment and software platform of the ROS robot, the experimental environment is built. On this basis, an improved YOLOv5-Lite object detection algorithm combining multi-scale attention and bounding box regression is proposed to form the ROS–YOLOv5–FleetEMA model, and the object detection function is integrated and deployed on the platform of the ROS robot. Through experimental analysis, relative to the conventional YOLOv5-Lite model, the ROS–YOLOv5–FleetEMA model enhanced the mAP @ 0.5 by 2.7%, reduced GFLOPS by 13.2%, and decreased the params by 15.1%; it has been proven that the ROS–YOLOv5–

FleetEMA model proposed in this paper can achieve near real-time object detection function. Compared with the traditional model, ROS–YOLOv5–FleetEMA shows significant advantages in a resource-constrained environment, including but not limited to high detection accuracy, small model size, low cost, and fast inference speed. These advantages give the ROS–YOLOv5–FleetEMA model an extremely high reference value and use value in practical applications. Although the ROS–YOLOv5–FleetEMA model proposed in this paper performs well in a specific experimental environment, its generalization ability for other types of datasets or practical application scenarios may be insufficient. In the future, we will study how to improve the generalization ability of the model further so that it can adapt to a wider range of application requirements.

Author Contributions: Conceptualization, H.W. and J.Z.; methodology, Z.S. and G.G.; software, Z.S. and G.G.; validation, H.W., C.L. and J.Z.; visualization, Z.X. and C.L.; writing—original draft, Z.S.; writing—review and editing, Z.S., G.G., Z.X. and H.W. All authors have read and agreed to the published version of the manuscript.

Funding: This research was funded by the "Jilin Province Science and Technology Development Plan Project, grant number YDZJ202201ZYTS549", the "Changchun Science and Technology Development Plan Project, grant number 21ZGM30", and the "Science and Technology Research Project of Education Department of Jilin Province, grant number JJKH20220597KJ.

Institutional Review Board Statement: Not applicable.

Informed Consent Statement: Not applicable.

Data Availability Statement: Publicly available datasets were used in this study. These data can be found here: https://github.com/rbgirshick/rcnn/issues/48 (accessed on 21 July 2024).

Acknowledgments: We would like to express our deepest gratitude to all those who have contributed to the completion of this research and the writing of this paper. Finally, special thanks to the Changchun University scholar climbing program for providing guidance on this research.

Conflicts of Interest: The authors declare no conflicts of interest.

References

1. Cao, Y. Analysis of the application of computer vision technology in automation. *China New Commun.* **2021**, *23*, 123–124.
2. Gu, Y.; Zong, X. A review of research on object detection based on deep learning. *Mod. Inf. Technol.* **2022**, *6*, 76–81.
3. Viola, P.; Jones, M. Rapid object detection using a boosted cascade of simple features. In Proceedings of the IEEE Computer Society Conference on Computer Vision and Pattern Recognition, Kauai, HI, USA, 8–14 December 2001; IEEE: New York, NY, USA, 2001; pp. 786–790.
4. Girshick, R.; Donahue, J.; Darrell, T.; Malik, J. Rich feature hierarchies for accurate object detection and semantic segmentation. In Proceedings of the IEEE Conference on Computer Vision and Pattern Recognition, Columbus, OH, USA, 23–28 June 2014; IEEE: New York, NY, USA, 2014; pp. 580–587.
5. Krizhevsky, A.; Sutskever, I.; Hinton, G.E. Imagenet classification with deep convolutional neural networks. *Commun. ACM* **2017**, *60*, 84–90. [CrossRef]
6. He, K.; Zhang, X.; Ren, S.; Sun, J. Spatial pyramid pooling in deep convolutional networks for visual recognition. *IEEE Trans. Pattern Anal. Mach. Intell.* **2015**, *37*, 1904–1916. [PubMed]
7. Zhao, W.; Fu, H.; Luk, W.; Yu, T.; Wang, S.; Feng, B.; Ma, Y.; Yang, G. F-CNN: An FPGA-based framework for training Convolutional Neural Networks. In Proceedings of the 2016 IEEE 27th International Conference on Application-Specific Systems, Architectures and Processors (ASAP), London, UK, 6–8 July 2016; IEEE: New York, NY, USA, 2016. [CrossRef]
8. Girshick, R. Fast r-cnn. In Proceedings of the IEEE International Conference on Computer Vision, Santiago, Chile, 7–13 December 2015; pp. 1440–1448.
9. Ren, S.; He, K.; Girshick, R.; Sun, J. Faster r-cnn: Towards real-time object detection with region proposal networks. In *IEEE Transactions on Pattern Analysis and Machine Intelligence*; IEEE: New York, NY, USA, 2017; Volume 39, pp. 1137–1149. [CrossRef]
10. Lin, T.Y.; Dollár, P.; Girshick, R.; He, K.; Hariharan, B.; Belongie, S. Feature Pyramid Networks for Object Detection. In Proceedings of the IEEE Conference on Computer Vision and Pattern Recognition (CVPR), Honolulu, HI, USA, 21–26 July 2017. [CrossRef]
11. Dai, J.; Li, Y.; He, K.; Sun, J. *R-FCN: Object Detection via Region-Based Fully Convolutional Networks*; Curran Associates Inc.: Red Hook, NY, USA, 2016. [CrossRef]
12. Qian, W. Research on Indoor Mobile Robot Target Detection and Location Grasping. Master's Thesis, Nanjing Forestry University, Nanjing, China, 2023.

13. Redmon, J.; Divvala, S.; Girshick, R.; Farhadi, A. You only Look Once: Unified, Real-time Object Detection. In Proceedings of the 2016 IEEE Conference on Computer Vision and Pattern Recognition (CVPR), Las Vegas, NV, USA, 27–30 June 2016; IEEE: New York, NY, USA, 2016; pp. 779–788.
14. Redmon, J.; Farhadi, A. YOLO9000: Better, Faster, Stronger. In Proceedings of the 2017 IEEE Conference on CVPR, Honolulu, HI, USA, 21–26 July 2017; IEEE: New York, NY, USA, 2017; pp. 6517–6525.
15. Redmon, J.; Farhadi, A. Yolov3: An incremental improvement. *arXiv* **2018**, arXiv:1804.02767.
16. Bochkovskiy, A.; Wang, C.Y.; Liao HY, M. Yolov4: Optimal speed and accuracy of object detection. *arXiv* **2020**, arXiv:2004.10934.
17. Wang, C.Y.; Liao, H.Y.M.; Wu, Y.H.; Chen, P.Y.; Hsieh, J.W.; Yeh, I.H. CSPNet: A New Backbone that can Enhance Learning Capability of CNN. In Proceedings of the 2020 IEEE/CVF Conference on Computer Vision and Pattern Recognition Workshops (CVPRW), Seattle, WA, USA, 14–19 June 2020; IEEE: New York, NY, USA, 2020. [CrossRef]
18. Ge, Y.; Qi, Y.; Meng, X. YOLOv5 Improved Lightweight Mask Face Detection. *Comput. Syst. Appl.* **2023**, *32*, 195–201. [CrossRef]
19. Lyu, Z.; Xu, Y.; Xie, Z. Detection of safety equipment for coal mine electric power personnel based on lightweight YOLOv5. *J. Heilongjiang Univ. Sci. Technol.* **2023**, *33*, 737–742.
20. Park, H.; Yoo, Y.; Seo, G.; Han, D.; Yun, S.; Kwak, N. C3: Concentrated-Comprehensive Convolution and its application to semantic segmentation. *arXiv* **2018**, arXiv:1812.04920. [CrossRef]
21. Ioffe, S.; Szegedy, C. Batch Normalization: Accelerating Deep Network Training by Reducing Internal Covariate Shift. In Proceedings of the International Conference on Machine Learning, Lille, France, 6–11 July 2015. [CrossRef]
22. Ma, N.; Zhang, X.; Zheng, H.T.; Sun, J. ShuffleNet V2: Practical Guidelines for Efficient CNN Architecture Design. In Proceedings of the European Conference on Computer Vision, Munich, Germany, 8–14 September 2018; Springer: Cham, Switzerland, 2018. [CrossRef]
23. Ouyang, D.; He, S.; Zhang, G.; Luo, M.; Guo, H.; Zhan, J.; Huang, Z. *Efficient Multi-Scale Attention Module with Cross-Spatial Learning*; Aerospace Science & Industry ShenZhen (Group) Co., Ltd.: Shenzhen, China, 2023.
24. Hajič, J., Jr.; Pecina, P. Detecting Noteheads in Handwritten Scores with ConvNets and Bounding Box Regression. *arXiv* **2017**, arXiv:1708.01806. [CrossRef]
25. Han, K.; Wang, Y.; Tian, Q.; Guo, J.; Xu, C.; Xu, C. GhostNet: More featuresfrom cheap operations. In Proceedings of the 2020 IEEE/CVF Conference on Computer Vision and Pattern Recognition, Seattle, WA, USA, 13–19 June 2020; IEEE: Seattle, WA, USA, 2020; pp. 1577–1586.
26. Support Government. Design and Implementation of Human Computer Interaction Interface for Seven Degree of Freedom Robotic Arm Based on ROS and Qt. Master's Thesis, China University of Petroleum (East China), Dongying, China, 2019. [CrossRef]
27. Hu, J.; Shen, L.; Albanie, S.; Sun, G.; Wu, E. Squeeze-and-Excitation Networks. *IEEE Trans. Pattern Anal. Mach. Intell.* **2019**, *42*, 2011–2023. [CrossRef] [PubMed]
28. Woo, S.; Park, J.; Lee, J.Y.; Kweon, I. *CBAM: Convolutional Block Attention Module*; Springer: Cham, Switzerland, 2018. [CrossRef]
29. Wang, Q.; Wu, B.; Zhu, P.; Li, P.; Zuo, W.; Hu, Q. ECA-Net: Efficient Channel Attention for Deep Convolutional Neural Networks. In Proceedings of the 2020 IEEE/CVF Conference on Computer Vision and Pattern Recognition (CVPR), Seattle, WA, USA, 13–19 June 2020; IEEE: New York, NY, USA, 2020. [CrossRef]
30. Hou, Q.; Zhou, D.; Feng, J. Coordinate Attention for Efficient Mobile Network Design. In Proceedings of the IEEE/CVF Conference on Computer Vision and Pattern Recognition, Nashville, TN, USA, 20–25 June 2021; IEEE: New York, NY, USA, 2021. [CrossRef]

applied sciences

MDPI

Article

Knowledge Embedding Relation Network for Small Data Defect Detection

Jinjia Ruan [1,†], Jin He [1,*,†], Yao Tong [1,*,†], Yuchuan Wang [1,†], Yinghao Fang [2] and Liang Qu [3]

1 China Waterborne Transport Research Institute, Beijing 100088, China; ruanjinjia@wti.ac.cn (J.R.); wych@wti.ac.cn (Y.W.)
2 Shandong Maritime Safety Administration, Qingdao 266002, China; fyh_msa@126.com
3 China National Offshore Oil Corporation, Tianjin 300459, China; quliang2@cnooc.com.cn
* Correspondence: hejin@wti.ac.cn (J.H.); tongyao@wti.ac.cn (Y.T.)
† These authors contributed equally to this work.

Abstract: In industrial vision, the lack of defect samples is one of the key constraints of depth vision quality inspection. This paper mainly studies defect detection under a small training set, trying to reduce the dependence of the model on defect samples by using normal samples. Therefore, we propose a Knowledge-Embedding Relational Network. We propose a Knowledge-Embedding Relational Network (KRN): firstly, unsupervised clustering and convolution features are used to model the knowledge of normal samples; at the same time, based on CNN feature extraction assisted by image segmentation, the conv feature is obtained from the backbone network; then, we build the relationship between knowledge and prediction samples through covariance, embed the knowledge, further mine the correlation using gram operation, normalize the power of the high-order features obtained by covariance, and finally send them to the prediction network. Our KRN has three attractive characteristics: (I) Knowledge Modeling uses the unsupervised clustering algorithm to statistically model the standard samples so as to reduce the dependence of the model on defect data. (II) Covariance-based Knowledge Embedding and the Gram Operation capture the second-order statistics of knowledge features and predicted image features to deeply mine the robust correlation. (III) Power Normalizing suppresses the burstiness of covariance module learning and the complexity of the feature space. KRN outperformed several advanced baselines in small training sets on the DAGM 2007, KSDD, and Steel datasets.

Keywords: defect detect; small training set; knowledge embedding relational network; gram operation

1. Introduction

Industrial quality inspection is one of the important application directions of computer vision in smart factories. However, in the industrial environment, defect samples are very scarce, and there may even be zero samples for some defect categories [1]. This makes small training sets a key constraint on the industrial implementation of many data-driven algorithms [2].

In the context of small training datasets, industry scholars have explored two main approaches: increasing the amount of data and reducing the dependency of algorithms on data. The former approach mainly generates new data through augmentation or introduces new data from other datasets [3,4], while the latter focuses on model improvement [5] and algorithm optimization [6] to enhance the feature extraction capabilities of small training sets. However, these methods from other machine learning applications in industrial manufacturing may not be directly applicable to industrial computer vision problems. For instance, in defect detection, defects may not be consistently present, or only a limited number of samples may be available over an extended period. Publicly available defect datasets also suffer from a scarcity of defect samples: the KolektorSDD dataset [7] contains only 52 defect samples out of 399 total, the AITEX dataset [8] has only 105 defect samples, and each category in the RSDDs dataset [9] has only 300 samples. This makes it challenging to effectively apply the aforementioned methods in these contexts.

Therefore, we believe that the small training set of industrial vision requires the introduction of external knowledge. Crucially, we noticed that there is an important difference between defect detection and target detection, that is, in addition to defect samples in the support set of defect detection, standard samples of products are also given, but target detection only has a small number of object samples [10].

Most of the existing automatic inspection equipment manufacturers do not use data-driven AI algorithms for automatic optical inspection, and their equipment generally still uses artificially designed features, although these thresholds test the engineer's ability to adjust parameters [11]. In fact, the deep learning algorithm can capture some latent features so that it far exceeds the ability of traditional algorithms in a single dataset, but in the actual industrial production environment, it cannot mine the correct features from limited samples, that is, the dataset External defect characteristics. The traditional "statistical modeling + similarity matching" algorithm can be more robust.

Inspired by this, we propose an idea to perform statistical modeling on standard samples as an auxiliary knowledge representation. This standard sample will participate in prediction and assist in identifying product quality inspection. Unlike other algorithms, we use standard samples as a priori knowledge rather than an input. Specifically, we introduce the Gram Matric for Knowledge Modeling. The Gram Matric has achieved outstanding results in the field of style transfer [12]. Some scholars have introduced it into few-shot classification and showed amazing results [13]. We believe that the covariance operation in the Gram Matric can deeply mine pixel-level feature correlations, which can also be considered to be related to texture features. Therefore, we introduce it into Knowledge Modeling and use it as an adjunct to enhance the identification of known and unknown defects.

The main contributions of this paper are as follows:

1. An aim at the lack of defect samples in industrial quality inspection scenarios, starting from external knowledge, using statistical modeling methods to build standard templates, using them as prior knowledge, and designing a defect detection network enhanced by prior knowledge;
2. In order to measure the difference between the standard sample and the predicted sample, as well as to embed this difference into the feature for subsequent head defect identification, a Knowledge-Embedding module based on self-attention was designed;
3. In order to obtain the relationship between features in the vector space and mine weak clues, we designed an eccentric covariance matrix to extract the characteristics of each dimension of the statistic, we automatically adjusted the unnecessary information in the extraction process avoiding interference from the cluttered background.
4. We demonstrated the effectiveness of this method on the public DAGM2007 [14], KolektorSDD [15], and Severstal Steel defect detection datasets.

2. Literature Review

Small Training Sets. Small training sets has always been a huge challenge in the application practice of deep machine vision. Even small training sets leads to a machine learning task of "decomposing the dataset into different meta tasks to understand the generalization ability of the model when the category changes"—otherwise known as Few-Shot Learning [16]. There are three methods of small training sets: data augmentation, model improvement, and algorithm optimization [17]: Data augmentation is used to expand the training data through various image methods to achieve the effect of increasing training samples, such as mixup [18], adding noise [19], and generating samples based on GANs [20]. In recent years, Pseudo-Labeling [21,22] has also become an effective method to improve performance points. Model improvement refers to adjusting the model structure to enhance the feature extraction ability [23]. The optimization algorithm is used to adjust learning strategies to improve algorithm performance. Semisupervised and unsupervised have also become a popular idea (to solve small training set problems) [24]. This paper focuses on introducing a priori knowledge and improving the model to reduce the dependence of the algorithm on defect data.

Active Shape Model. The statistical model of the PCB standard board has a strong positive impact on defect detection. The statistical shape modeling technology was proposed by Cootes in his paper [25] in 1995. It is a deformable model in computer vision, which is used to model the shape in the image. This method only needs to establish a flexible mathematical model and only needs to compare each time. Using this method, the debugging efficiency of the AOI is accelerated, and the misjudgment rate is reduced. Inspired by this, this paper encodes the standard image through CNN, makes statistical analysis on the standard samples by using the clustering algorithm, obtains representative standard samples, and constructs the standard template as the representation of a priori knowledge.

Self-Attention Modules. They have been successfully applied in NLP [26] and physical system modeling [27]. The self-attention mechanism can capture the relationship between the original sentence and the target sentence in natural language processing, and replace the recurrent neural network with an attention model, so as to realize parallel implementation and more efficient learning. These works inspire us to deduce the variant of knowledge embedding based on correlation mining. We converted the original elements from words to conv features and employed the knowledge model of the predicted image. We used this mechanism to establish the knowledge embedding method in the feature mapping from the low dimension to the high dimension.

Gram Matrix. In fact, it can be regarded as an eccentric covariance matrix between features, that is, a covariance matrix without mean subtraction. Second-order statistics have been studied in the context of texture recognition through so-called regional covariance descriptors (RCDs), which were further applied to object class recognition [28]. Co-occurrence patterns can also be used in the CNN setting. A recent approach [29] extracted feature vectors at two separate locations in a feature map and performed an outer product to form a CNN co-occurrence layer. Higher-order statistics have also been used for fine-grained image classification [30] and domain adaptation [31]. SoSN utilizes second-order information and power normalization for end-to-end training with one- or few-shot learning. Based on the second-order statistics applied to these matrics, we designed a multi-relational feature descriptor that captures deep relationships between proposals before being passed to the classification network for defect identification.

3. Research Methods

Below, we introduce our deep template matching defect detector network and then describe its individual components.

3.1. Overview

In this paper, this method operates on the so-called small training set defect detection, which is essentially a classification task. However, in different scenarios, defects are detected, and segmentation is also a task requirement. Taking classification as the main goal, we evaluate the segmentation and detection of defects.

Different from some defect detection schemes that simply add negative samples, we use standard samples as a priori knowledge to identify defects by mining feature relationships. Our Knowledge-Embedding Relational Network (KRN) consists of (i) an encoding network, (ii) Knowledge Modeling, (iii) Knowledge Embedding, (iv) a Gram Operation, and (v) a Prediction Network. Figure 1 shows an example of an architecture that supports images.

The role of the Encoding Network is to generate image-level convolutional feature vectors (descriptors), and our Encoding Network includes the segmentation part. The task of the Knowledge Modeling part is to perform statistical modeling on multiple standard samples in order to obtain a knowledge representation that can assist in enhancing defect detection. The knowledge association embedding module is an operation of mining the relationship between prior knowledge and predicted samples, aiming to promote the fusion of prior knowledge and predicted samples. The task of the Gram Operation is to use the Gram Matrix to mine the latent relationship between each feature vector so as to make the defect salient. Finally, the Predictive Network learns and recognizes this knowledge-embedded relation mining feature.

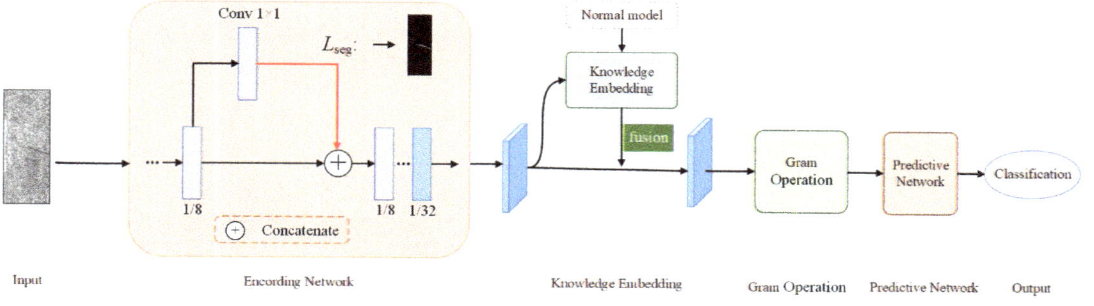

Figure 1. The architecture of general Defect Detection models. (1) Encoding Network; (2) Knowledge Embedding consist of knowledge model, correlation fusion module, and Gram Operation; (3) Predictive Network.

3.2. Encoding Network

The feature encoding network is responsible for generating convolutional feature vectors, which serve as image descriptors. To address the challenges of sample concentration, high resolution, and small target scenes in industrial visual defect detection tasks, this paper utilizes a convolutional neural network architecture based on ESDN [32]. Specifically, we employed the Segmentation Network and Decision Network, which perform downsampling by a factor of 32, as the feature encoding network. It is important to note that the segmentation component was used as an auxiliary module for feature extraction.

The Encoding Network can be described as $f : (\mathbb{R}^{W\times H}; \mathbb{R}^{|\mathcal{F}|}) \rightarrow \mathbb{R}^{W\times H}$, where W and H represent the width and height of the input image.The Encoding Network f is a convolutional neural network specifically designed for feature extraction in industrial visual defect detection tasks. It takes an input image of size $W \times H$ and produces a feature map of the same spatial dimensions. This network includes multiple convolutional layers, which downsample the input by a factor of 8. After downsampling, the output is split into two branches: one branch undergoes segmentation using a 1×1 convolution, and the result is concatenated with the original downsampled feature map. This architecture is optimized to retain defect details by operating at a middle scale, balancing the trade-off between computational efficiency and the preservation of important features.

3.3. Correlation Knowledge Embedding

3.3.1. Knowledge Modeling

Defect detection is a small training sets recognition task, and a large number of normal samples can be used as a reference. Therefore, we specially designed the Knowledge Modeling module, which aims to perform knowledge mining on normal samples for the reference of defect recognition.

Specifically, we used resnet50 to map standard samples to high-dimensional features and convert them into tensors. The weight is trained on Imagenet and has extensive classification ability. Then, n representative images are selected from a large number of standard samples by a clustering algorithm, and then a multi-dimensional image composed of overlapping standard images is defined as $X_{norm} \in \mathbb{R}^{W\times H}$, that is, the constructed statistical knowledge.

3.3.2. Knowledge Embedding

Then, we designed a knowledge embedding method based on self-attention, as shown in Figure 2.

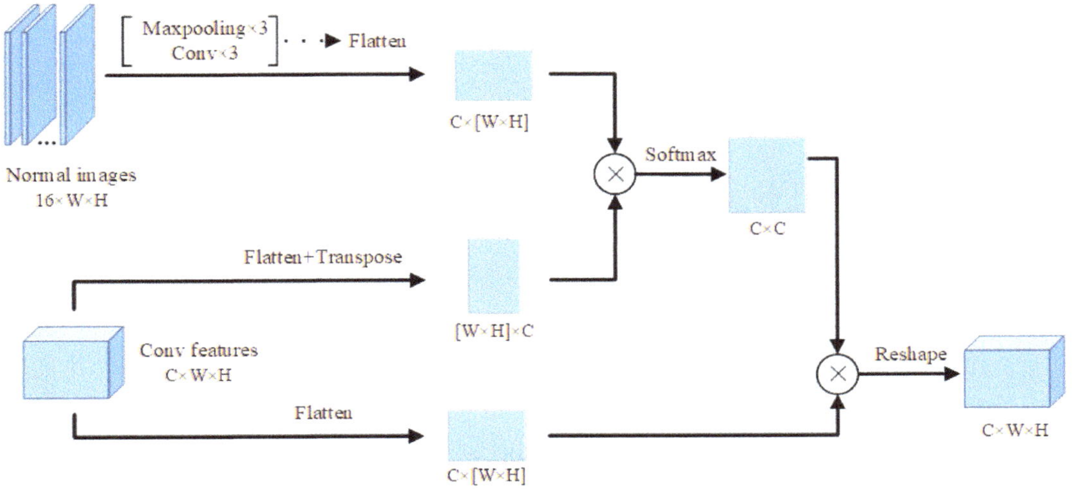

Figure 2. Knowledge mining based on embedded relation module. The input is the conv characteristics of the predicted samples and the prior knowledge processed into tensors. The correlation between the two is captured with the help of covariance operation and then fused in the form of attention.

The simplest knowledge fusion is concat and add operations, but we hope that this fusion can excavate the correlation between the two to a certain extent for fusion. Referring to the self-attention mechanism, we designed a knowledge embedding method based on relationship mining. In our Knowledge Embedding module, the input consists of conv feature Φ_{pre} and knowledge feature Φ_{norm}; Φ_{pre} comes from the last convolution output feature of the Encoding Network. Φ_{norm} is the shallow tensor obtained by X_{norm} through three maxpoling and three groups of convolution sampling, as shown in formula (1).

$$\Phi_{norm} = f_{norm}(X_{norm}; F), \Phi_{norm} \in \mathbb{R}^{K\times N} \tag{1}$$

where F are the parameters to learn of three convolution layers in the Knowledge Model.

A dot product is performed between the conv features and knowledge features to obtain their correlation. A softmax function is applied to obtain the weights on the values. Given matrices ϕ_{norm} (by flattening Φ_{norm}) and matrices ϕ_{pre} (by flattening Φ_{norm}), their correlation is computed as follows:

$$M = s(\Phi_{norm}, \Phi_{pre}) = softmax\left(\phi_{norm}\phi_{pre}^{T}\right) \tag{2}$$

This correlation will be embedded into the original conv feature through multiplication. The output is computed as

$$\phi_{out} = mul(M, \phi_{pre}) = softmax\left(dot\left(\phi_{norm}, \phi_{pre}^{T}\right)\right)\phi_{pre} \tag{3}$$

Finally, the feature matrix after Knowledge Embedding will be reshape as a feature maps.

In addition, after visiting an electronic factory, we speculated that adding the standard template would help the consumer electronics industry with some wrong parts and defects. They had no appearance damage defects, but the welded components were inconsistent with the design drawings. Having been limited by this dataset, we may test it in future practice.

3.4. Gram Operation

The Gram Matrx is an operation to deeply mine the correlation between features, as shown in Figure 3. We used it as a feature mining tool for defect textures and normal textures. Its input is a feature vector from Knowledge Embedding, which we define as $\Phi = \{\phi_n\}_{n\in\mathcal{N}}$. Then, we used $\phi\phi^T = \uparrow \otimes_2 \phi$ to denote the covariance operation of the eigenvectors. Taking Φ for example,

$$\Psi(\Phi_n) = \frac{1}{N}\sum_{n\in\mathcal{N}_s} \uparrow \otimes_r \phi_n = \Psi\left(\{\phi_n\}_{n\in\mathcal{N}_s}\right) = \frac{1}{N}\sum_{n\in\mathcal{N}_s} \phi_n\phi_n^T \quad (4)$$

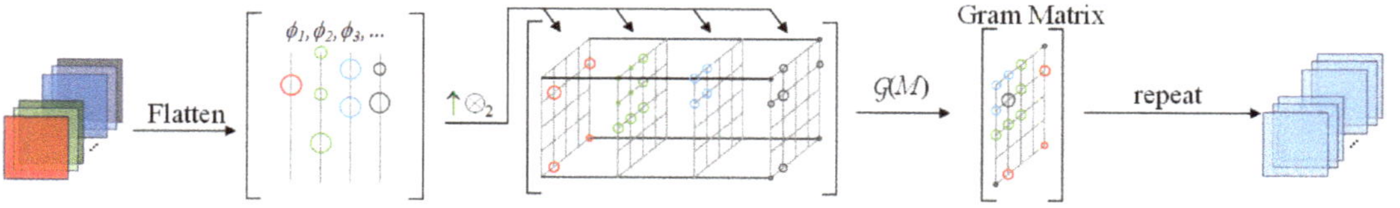

Figure 3. Gram Operation module. We flatten conv features into feature vectors, capture second-order features by covariance operation, and then send them to PN. Finally, self replication is carried out for the subsequent diversity and fusion promotion operation.

KE and GO capture feature correlation through covariance, which essentially introduces second-order statistics. Second-order statistics have to deal with the so-called burstiness, which is "the property that a given visual element appears more times in an image than a statistically independent model would predict". Power Normalization [13] is known to suppress this burstiness and has been extensively studied and evaluated in the context of Bag-of-Words and Few-Shot Learning. Therefore, we adopted SigmE PN which is defined as

$$\mathcal{G}_{SigmE}(M,\eta) = \frac{2}{1+e^{\frac{-\eta' M}{Tr(M)+\lambda}}} - 1 \quad (5)$$

where $1 \leq \eta \approx N$ interpolates between counting and detection, $\lambda \approx le^{-6}$ is a regularization constant, and the trace $Tr()$ stops the diagonal from exceeding 1.

After calculating the Gram Matrix, it is replicated 16 times in the channel dimension to ensure compatibility with the subsequent predict network layers, which are designed to process a specific number of input channels. While using a single-channeled Gram Matrix could reduce computational redundancy, this would require significant architectural changes to the predict network, potentially affecting its performance and stability. The replication maintains the continuity and integrity of the convolutional bottleneck structure without altering the network's existing architecture.

3.5. Predictive Network

The function of the Prediction Network is to mine and judge the features of the Knowledge Embedding, as well as realize the detection of the target image. We did not directly predict the features of the Knowledge Embedding but used three Conv Blocks: the specific parameters are two Conv 1×1 with channel 32 and one Conv 3×3 with channel 16, implemented using convolution operations to facilitate feature mining between standard templates and predicted samples.

3.6. Loss Function

The loss function of the Knowledge-Embedding Relation Network (DKER) consists of two parts: the loss L_{seg} for segmentation that is assisted and the classification loss L_{cls}. The total loss can be denoted as $L = \lambda Lseg + \delta(1 - \lambda)L_{cls}(M)$, where λ is a simple linear function, and δ is a weight coefficient with a small value. The detailed sets were defined as in [32].

4. Experiments

Below, we experimentally demonstrate the merits of our Knowledge-Embedding Relation Network. Our method was mainly evaluated on the DAGM2007, KolektorSDD, and Severstal Steel datasets. We compared with other advanced algorithms, designed a small training sets test, and conducted ablation experiments.

4.1. Datasets

DAGM2007 contains texture data of 10 categories, and each category contains 1000 negative samples and 150 positive samples saved in grayscale 8-bit PNG format. The training set and test set of each category were allocated in a proportion of 1:1, and the size was a 512 × 512 image. In addition, we explored small training sets scenarios of 5, 10, 15, 20, and 25 positive samples in proportion. It should be pointed out that the standard sample modeling in all training was established from the standard samples of the complete dataset.

The KolektorSDD dataset includes eight non-overlapping images collected from each commutator surface of 50 defective electronic commutators, and a total of 399 images were obtained, including 52 defective images and 347 defect-free images. All data settings refer to dagm, except that the image size was 1408 × 512 for smaller datasets. In addition, our ablation experiment was evaluated on the KSDD dataset with five positive samples.

The Severstal Steel dataset is from the Kaggle Challenge, which contains 12,568 images and involves four kinds of defects. In the effectiveness demonstration of this method, we adopted the scheme of 1000 positive samples, and the size of the input image was 256 × 1600. In the small training sets scenario, our data settings were the same as above, but the number of test sets was not changed (in order to be more consistent with the real scenario).

4.2. Implementation Details

All codes were implemented in PyTorch. All experiments were tested in the PyTorch framework under the Ubuntu system, and two Titan Xs were used for GPU acceleration. For the learning rate, we followed the learning scheme [32], and DAGM adopted $LR = 0.01$ and $\delta = 1$. KSDD adopted $LR = 0.5$ and $\delta = 0.01$, as well as $LR = 0.1$ and $\delta = 0.1$. However, for the number of learning iterations, due to the introduction of high-order moments, the high-dimensional mapping of features dragged down the convergence speed to a certain extent at the beginning of training, so we adjusted the number of iterations of the experiment: in the small training sets of 5, 10, 15, 20, and 25 positive samples, we trained 350, 190, 170, 150, and 140 epochs, respectively. In the complete experiment, the three data trained 150 epochs.

We conducted less exploration on training tricks and paid more attention to less sample training. In the following experiments, we compare the research results with several advanced methods, and we report the commonly used matrices for KRN, such as AP, FP, and FN.

4.3. Comparison with the State of the Art

4.3.1. DAGM2007

The proposed KRN was evaluated on the DAGM 2007 dataset, and the obtained true positive rate (TPR) and true negative rate (TNR) are shown in Table 1. Our method achieved 100% TPR and TNR on all folds, which means its completely solved this dataset. Practically, the ESDN had achieved this goal before that. Some other explorations also achieved high scores, such as Racki et al. [33], who obtained nine 100% outcomes and a 98.5% in a ten fold, and Kim et al. [34], who obtained 100%, except fold 1 and fold 4 Dagm, as a classic dataset of material texture, whih has sufficient data samples. We tested it above to prove that the KRN guarantees a high score on this complete dataset. We visualized some results in Figure 4.

Table 1. mAP on four methods (DAGM, 150 positive samples).

Surface	Our		ESDN		Racki et al.		Kim et al.		Scholz et al. [35]	
	TPR	TNR	TPR	TNR	TPR	TNR	TPR	TNR	TPR	TNR
1	100	100	100	100	100	98.8	99.8	100	99.7	99.4
2	100	100	100	100	100	99.8	100	100	80.0	94.3
3	100	100	100	100	100	96.3	100	100	100	99.5
4	100	100	100	100	98.5	99.8	99.9	100	96.1	92.5
5	100	100	100	100	100	100	100	100	96.1	96.9
6	100	100	100	100	100	100	100	100	96.1	100
7	100	100	100	100	100	100	-	-	-	-
8	100	100	100	100	100	100	-	-	-	-
9	100	100	100	100	100	99.9	-	-	-	-
10	100	100	100	100	100	100	-	-	-	-

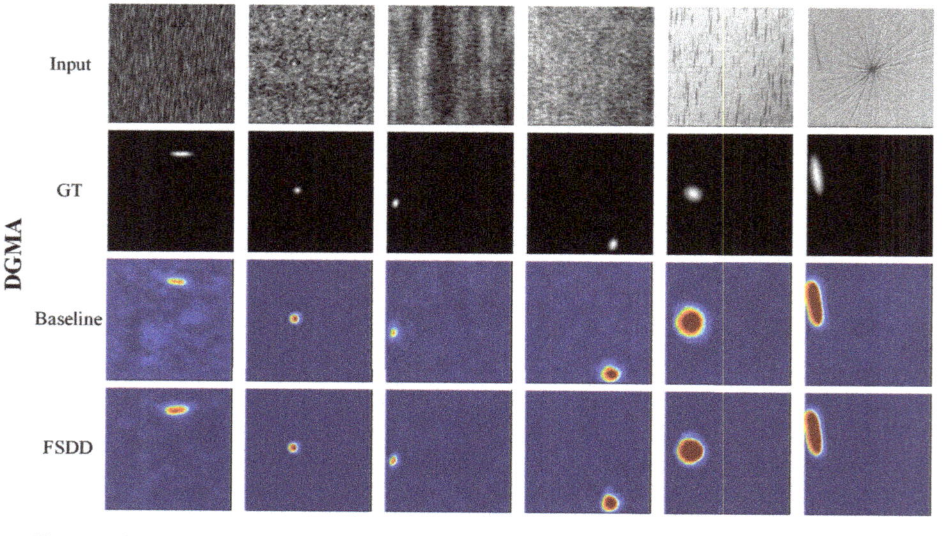

Figure 4. *Cont.*

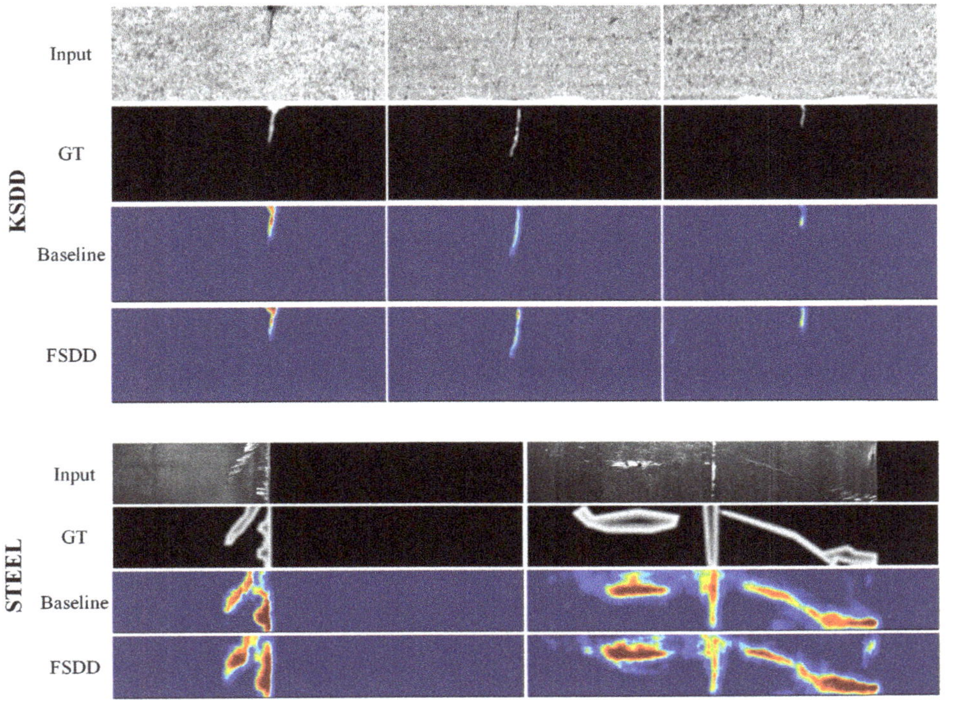

Figure 4. Examples of images, defects, and detections with segmentation output from the DAGM (**top**), KolektorSDD (**middle**), and Steel (**bottom**) datasets.

4.3.2. KolektorSDD

The proposed KRN is compared with the ESDN, SDN, and EfficientNet in Table 2. The KSDD is a totally industrial few-shot dataset with only 53 positive samples. Its author defines it as a classification problem and especially adds small-scale segmentation annotation as an auxiliary. Assisted by segmentation, the SDN obtained a higher score of 99.00%. After a series of optimizations such as dynamic balance loss and gradient adjustment, the ESDN became an end-to-end defect detection model, and it achieved a 99.49% AP and 1 + 2 (FP + FN) in our experiments. According to Table 2, Our KRN achieved a further performance of 100.00% AP and realized 0 + 0 (FP + FN).

Table 2. mAP on four methods (KSDD, 33 positive samples).

Method	AP/%	FP + FN
EfficientNet	-	-
SDN	99.00	1 + 0
ESDN	99.49	1 + 2
Ours	100.00	0 + 0

4.3.3. Severstal Steel

Table 3 compares our KRN with the ESDN, SDN, and EfficientNet on Severstal steel for 1000 positive samples. As shown in the table, in contrast, the KRN was the best among all methods. Its performance on the AP was 1.18%, 7.38%, and 8.17% higher than the ESDN, SDN, and EfficientNet, respectively.

Table 3. mAP on four methods (Steel, 1000 positive samples).

Method	AP/%	FP + FN
EfficientNet	91.56	-
SDN	92.35	-
ESDN	98.45	68 + 85
Ours	99.73	40 + 32

4.4. Ablation Study

Below we analyze the effectiveness of each component of the proposed KRN approach. We designed six groups of experimental variants (except for the ablation experiment of each component), including variants used by all components and variants not used by all components. The following ablation studies were based on the KSDD dataset with the five positive samples setting. For some components, we further designed comparative experiments for in-depth analysis. For example, the Knowledge Embedding part also adopted add, concat, and mat operations.

Knowledge Modeling (KM): We conducted KM on flawless samples. In this part of the ablation experiment, we changed the KM to predict the conv feature of the image to simulate the effect of no KM. The experimental results show that KM has a great impact on the KRN (97.39% vs 94.88%). For KSDD settings, it shows that without statistical modeling in Table 4, the AP scores of our KRN on fold 0 and fold 1 decreased by 6.2% and 3.28%, respectively. The analysis shows that KM brings additional knowledge and enhances the recognition performance of the model.

Table 4. mAP on four methods (ablation study on KSDD Dataset with 5 positive samples).

Fold 0	Fold 1	Fold 2	Mean	Knowledge Model	Knowledge Embedding	Gram Operation	Power Normalizing
94.90	90.01	98.47	94.46				
93.18	94.70	96.75	94.88		√	√	√
99.43	91.58	**8.47**	96.49	√		√	√
94.28	95.00	96.08	95.12	√	√		√
96.84	91.21	96.31	94.79	√	√	√	
99.38	**97.98**	94.80	**97.39**	√	√	√	√

Knowledge Embedding (KE): In the ablation experiment of KE, our KRN performed the variant (97.39% vs. 96.49%), which levers the regular concat fusion in Table 4. concat had the best AP value on fold 0 and fold 2, but fold 1 had only a 91.58% AP. The KE module has the ability to capture knowledge features and predict the relationship between sample features in design, because the module has been significantly enhanced in the defect identification of fold 1 (97.39% vs. 91.58%).

We additionally analyzed the impact of concat, add, and mul operations on KE. The experimental results are shown in Table 5. Among them, the add operation was the roughest for feature fusion through simple addition, with the AP score being the lowest, which was 95.26%. The concat operation retained the original features and knowledge features, which was 1.23% higher than add. Mul enlarged the local difference, and the multiplication between the conv feature and knowledge feature was conducive to mining the relationship between the two features; its ap score was close to KE. Our KE not only retains the original features, but also constructs the relationship between the quasi-sample and the predicted sample, deeply excavates the potential differences, and shows the best performance (97.39% vs. 95.26%, 96.49%, and 97.26%).

Gram Operation (GO): The GO is a supplement to KE, mining the relationship between features after KE. As shown in Table 4, the performance of the GO defect detection algorithm had a 2.27% drop (97.39% vs. 95.12%). GO mines the relationship between features of the final total features. Based on the experimental results of three folds (94.28%,

95.00%, and 96.08%), the GO was one of the main components to improve the performance of the model. The results in fold 0 show that its contribution to the KRN is second only to KM.

Table 5. mAP on four variants of KE (KSDD, 5 positive samples).

Method	AP/%	FP + FN
Add	95.26	2 + 1
Concat	96.49	1 + 1
Mul	97.26	1 + 1
KE	97.39	0 + 1

Power Normalizing (PN): We also analyzed the improvement brought by the PN, which considers a Burst suppressor on second-order statistics. It can be seen from the previous experiments that the KE and GO introduced negative effects. The analysis shows that it was due to the burst of high-order statistics. In Table 6, it can be seen that after adding the PN operation, the negative effects brought by KE and GO were suppressed. The AP on fold 0 and fold 1 increased by 2.54% and 6.77%, respectively, and the average AP increased by 2.60%.

Table 6. mAP on Two PNs (KSDD, 5 positive samples).

Method	AP/%	FP + FN
None	94.79	2 + 2
AsinhE	96.89	0 + 2
SigmE	97.39	0 + 1

In particular, we explored two different PN strategies: Asinhe and SigmE. Without any power normalization, the AP score was only 94.79%, and FP + FN was 2 + 2. After adding Power Normalization, the performance was optimized, the false detection and missed detection were reduced from 4 to 2, and the AP increased to 96.89% and 97.39%, respectively. Our knowledge model relationship detector with sigma pooling is beneficial for small training sets defect detection.

In summary, KM, the GO, and PN have a significant impact on the three folds. Among them, KM and GO have a greater gain on fold 0, KE and PN have a greater impact on fold 1, and Ke seems to have a negative effect on fold 2. KM introduces external knowledge, KE embeds and fuses the knowledge, and GO and PN further promote the integration. The four modules cooperate with each other to form gain, mine potential features, and improve the performance of the small training sets defect detection model.

4.5. Small Training Sets

We paid special attention to the scene, where it is difficult to obtain defect samples in the industrial scene, that is, a small number of positive samples. We used positive samples of 5, 10, 15, 20, and 25 for each dataset. The results are shown in Table 7.

For the DAGM2007 dataset, originally based on 150 positive samples, EfficientNet and ESDN could have good performance, but when the number of positive samples decreased, the detection performance of all algorithms showed different decline, as shown in the figure. The details of the decline can be seen in Table 7. When the number of positive samples decreased from 150 to 25 and then to 5, the AP value of the baseline was 100%–90.11%–82.39%; the AP value of the KRN 100%–99.11%–82.39%. When the training samples were sufficient, their AP scores were very high. When the number of positive samples decreased to 25, ESDN decreased by 3%, and KRN decreased by 1.47%; when the number of positive samples turned to five, the AP of the baseline decreased significantly (14%), and the KRN performed better, which only decreased by 9%.

For the KSDD dataset, all three methods achieved good scores. After analysis, we believe that this is because KSDD itself is a small dataset, and the test set is not complex, which makes it possible to obtain a good recognizer with only a small number of samples. However, five positive samples still brought differences to the performance of the various methods. According to Table 7, when the number of positive samples was greater than 20, the AP of each of the three methods was close to 100%. However, with the decrease in the number of positive samples to five, the baseline decreased by 4.18%, while the KRN was the least affected by the standard template KE and relationship mining module, which was only 2.65%

Regarding the performance for the Steel dataset, like DAGM, all methods were greatly reduced: in the scenario of five positive samples, the AP of the baseline decreased to 58.45%, and there were false detection and missed detection values of 490 + 25; the AP performance of the KRN was the best, which was 63.28%, and the false detection and missing detection came out to only 467 + 34. Although these benefits are lower than the DAGM and KSDD, they are consistent and significant considering that the Steel dataset is more challenging in terms of complexity and dataset size.

Table 7. mAP on Three Methods (DAGM, KSDD, STEEL).

Model	Dataset	Num of Positive Samples									
		5		10		15		20		25	
		AP/%	FP + FN	AP/%	FP + FN	AP/%	FP + FN	AP/%	FP + FN	AP/%	FP + FN
baseline	DAGM	85.58	43 + 12	91.02	43 + 5	98.09	2 + 3	99.27	2 + 2	99.18	1 + 2
Our		90.12	23 + 16	96.23	3 + 7	98.04	2 + 3	99.65	1 + 1	99.11	1 + 1
baseline	KSDD	95.82	2 + 2	97.25	2 + 2	97.96	1 + 2	98.74	1 + 1	99.78	0 + 1
Our		97.35	1 + 2	97.84	1 + 2	99.15	1 + 1	99.29	0 + 1	100.00	0 + 0
baseline	STEEL	58.45	490 + 25	54.27	508 + 18	64.81	334 + 65	62.87	419 + 56	67.04	401 + 61
Our		63.28	467 + 31	60.13	438 + 53	65.57	346 + 70	68.13	349 + 48	75.46	315 + 83

We visualized the experimental results,as shown in Figure 5. The AP and FP + FN increased by varying degrees with the increase in positive samples. It reveals the generalization fragility of the baseline under the small training sets: without adequate training images, it detected poorly defected from input images. In contrast, for our KRN, the KE and GO demonstrated superior performance on small training sets defect detection.

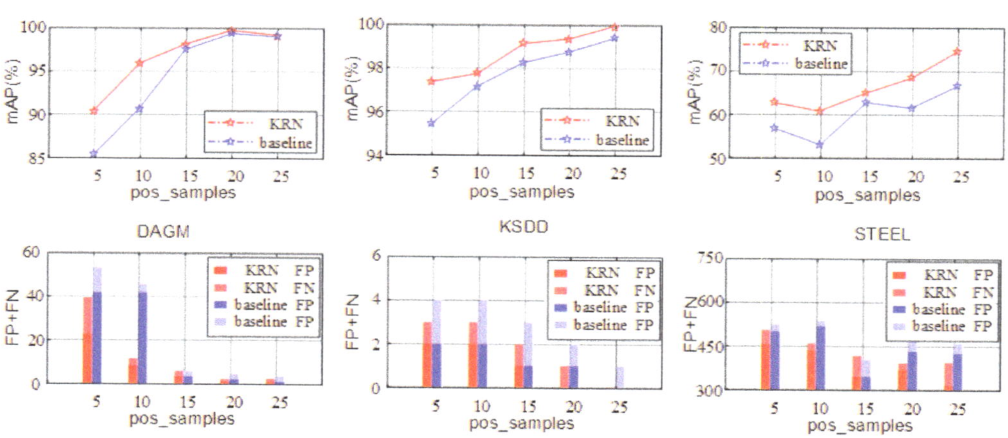

Figure 5. Smaller training set size results of DAGM, KSDD, Steel. The three figures above are the change curve of map with the number of positive samples, and the three figures below are the corresponding FP + FN.

4.6. Visualization of Detection Results

The following shows some test images on the DAGM, KSDD, Steel, and other datasets. We output the segmentation results, including the original image, ground truth, baseline, and KRN. Our model is more classification-based , and segmentation is the auxiliary part of conv feature extraction. Therefore, segmentation only needs shallow segmentation in the small and medium scale. Based on the visual results, the defect part in the segmented image will be larger than ground truth, which is normal. In addition, from Figure 5, in "the first sample of DAGM, the third sample of KSDD and the second sample of Steel", it can also be seen that the difference was amplified by the KRN after introducing the second-order moment of covariance. But at the same time, the higher-order feature increased the burst (the noise in the upper right of the second sample of Steel).

5. Conclusions

In this paper, we proposed a Knowledge-Embedding Relation Network (KRN) for the small training sets Defect Detection to address few-shot defect detection. Our model extends the ESDN through embedding standard templates and second-order statistics into CNN features of segmentation excitation. The standard template provides external knowledge for defect samples, while KE and GO provide high-latitude potential relationship features. In order to demonstrate the effectiveness of the KRN, we have conducted extensive quantitative and qualitative experiments on several datasets. In particular, we simulated the industrial detection scene of small training sets and carried out relevant experiments.

Author Contributions: Resources, Y.W.; Data curation, Y.T., Y.F. and L.Q.; Writing—original draft, J.R.; Writing—review & editing, J.H. All authors have read and agreed to the published version of the manuscript.

Funding: This work was supported in part by the National Key Research and Development Program of China, No. 2023YFC3107903, and in part by the National Key Research and Development Program of China, No. 2023YFB4302302.

Institutional Review Board Statement: Not applicable.

Informed Consent Statement: Not applicable.

Data Availability Statement: Data available in a publicly accessible repository. The DAGM2007 dataset presented in this study are available at [https://www.kaggle.com/datasets/mhskjelvareid/dagm-2007-competition-dataset-optical-inspection, accessed on 4 September 2024], the KolektorSDD dataset presented in this study are available at [https://www.vicos.si/resources/kolektorsdd/, accessed on 4 September 2024]. No new data were created or analyzed in this study.

Conflicts of Interest: Author Liang Qu was employed by the China National Offshore Oil Corporation. The remaining authors declare that the research was conducted in the absence of any commercial or financial relationships that could be construed as a potential conflict of interest.

References

1. Bao, Y.; Song, K.; Liu, J.; Wang, Y.; Yan, Y.; Yu, H.; Li, X. Triplet-graph reasoning network for few-shot metal generic surface defect segmentation. *IEEE Trans. Instrum. Meas.* **2021**, *70*, 5011111. [CrossRef]
2. Yu, W.; Zhang, Y.; Shi, H. Surface Defect Inspection Under a Small Training Set Condition. In Proceedings of the International Conference on Intelligent Robotics and Applications, Shenyang, China, 8–11 August 2019; pp. 517–528.
3. Saha, S.; Sheikh, N. Ultrasound image classification using ACGAN with small training dataset. In Proceedings of the International Symposium on Signal and Image Processing, Kolkata, India, 18–19 March 2020; pp. 85–93.
4. Si, C.; Zhang, Z.; Qi, F.; Liu, Z.; Wang, Y.; Liu, Q.; Sun, M. Better Robustness by More Coverage: Adversarial and Mixup Data Augmentation for Robust Finetuning. In *Findings of the Association for Computational Linguistics: ACL-IJCNLP 2021*; Association for Computational Linguistics: Stroudsburg, PA, USA, 2021; pp. 1569–1576.
5. Hsu, M.-J.; Chien, Y.-H.; Wang, W.-Y.; Hsu, C.-C. A convolutional fuzzy neural network architecture for object classification with small training database. *Int. J. Fuzzy Syst.* **2020**, *22*, 1–10. [CrossRef]

6. Wu, Y.; Lin, Y.; Dong, X.; Yan, Y.; Ouyang, W.; Yang, Y. Exploit the unknown gradually: One-shot video-based person re-identification by stepwise learning. In Proceedings of the IEEE Conference on Computer Vision and Pattern Recognition, Salt Lake City, UT, USA, 18–23 June 2018; pp. 5177–5186.
7. Tabernik, D.; Šela, S.; Skvarč, J.; Skočaj, D. Segmentation-based deep-learning approach for surface-defect detection. *J. Intell. Manuf.* **2020**, *31*, 759–776. [CrossRef]
8. Silvestre-Blanes, J.; Albero, T.; Miralles, I.; Pérez-Llorens, R.; Moreno, J. A public fabric database for defect detection methods and results. *Autex Res. J.* **2019**, *19*, 363–374. [CrossRef]
9. Gan, J.; Li, Q.; Wang, J.; Yu, H. A hierarchical extractor-based visual rail surface inspection system. *IEEE Sens. J.* **2017**, *17*, 7935–7944. [CrossRef]
10. Malamas, E.N.; Petrakis, E.G.M.; Zervakis, M.; Petit, L.; Legat, J.-D. A survey on industrial vision systems, applications and tools. *Image Vis. Comput.* **2003**, *21*, 171–188. [CrossRef]
11. Abd Al Rahman, M.; Mousavi, A. A review and analysis of automatic optical inspection and quality monitoring methods in electronics industry. *IEEE Access* **2020**, *8*, 183192–183271.
12. Li, Y.; Wang, N.; Liu, J.; Hou, X. Demystifying neural style transfer. *arXiv* **2017**, arXiv:1701.01036.
13. Zhang, H.; Koniusz, P. Power normalizing second-order similarity network for few-shot learning. In Proceedings of the 2019 IEEE Winter Conference on Applications of Computer Vision (WACV), Waikoloa, HI, USA, 7–11 January 2019; pp. 1185–1193.
14. Jager, M.; Knoll, C.; Hamprecht, F.A. Weakly supervised learning of a classifier for unusual event detection. *IEEE Trans. Image Process.* **2008**, *17*, 1700–1708. [CrossRef] [PubMed]
15. Ghatnekar, S. Use Machine Learning to Detect Defects on the Steel Surface. 2018. Available online: https://insiders.intel.com/projects/using-machine-learning-to-detect-defects-on-the-steel-surface (accessed on 2 September 2024).
16. Wang, W.; Zheng, V.W.; Yu, H.; Miao, C. A survey of zero-shot learning: Settings, methods, and applications. *ACM Trans. Intell. Syst. Technol. (TIST)* **2019**, *10*, 1–37. [CrossRef]
17. Wang, Y.; Yao, Q.; Kwok, J.T.; Ni, L.M. Generalizing from a few examples: A survey on few-shot learning. *ACM Comput. Surv. (CSUR)* **2020**, *53*, 1–34. [CrossRef]
18. Fu, Y.; Fu, Y.; Jiang, Y.-G. Meta-FDMixup: Cross-Domain Few-Shot Learning Guided by Labeled Target Data. In Proceedings of the Proceedings of the 29th ACM International Conference on Multimedia, Virtual, 20–24 October 2021; pp. 5326–5334.
19. Shorten, C.; Khoshgoftaar, T.M. A survey on image data augmentation for deep learning. *J. Big Data* **2019**, *6*, 60. [CrossRef]
20. Zhang, R.; Che, T.; Ghahramani, Z.; Bengio, Y.; Song, Y. Metagan: An adversarial approach to few-shot learning. *Adv. Neural Inf. Process. Syst.* **2018**, *31*.
21. Renz, K.; Stache, N.C.; Fox, N.; Varol, G.; Albanie, S. Sign Segmentation with Changepoint-Modulated Pseudo-Labelling. In Proceedings of the 2021 IEEE/CVF Conference on Computer Vision and Pattern Recognition Workshops (CVPRW), Nashville, TN, USA, 19–25 June 2021.
22. Lau, S.L.; Lew, J.; Ho, C.C.; Su, S. Exploratory Investigation on a Naive Pseudo-labelling Technique for Liquid Droplet Images Detection using Semi-supervised Learning. In Proceedings of the 2021 IEEE International Conference on Computing (ICOCO), Kuala Lumpur, Malaysia, 17–19 November 2021; pp. 353–359.
23. Koch, G.; Zemel, R.; Salakhutdinov, R. Siamese neural networks for one-shot image recognition. In Proceedings of the ICML Deep Learning Workshop, Lille, France, 6–11 July 2015; Volume 2.
24. Božič, J.; Tabernik, D.; Skočaj, D. Mixed supervision for surface-defect detection: From weakly to fully supervised learning. *Comput. Ind.* **2021**, *129*, 103459. [CrossRef]
25. Cootes, T.F.; Taylor, C.J.; Cooper, D.H.; Graham, J. Active shape models-their training and application. *Comput. Vis. Image Underst.* **1995**, *61*, 38–59. [CrossRef]
26. Vaswani, A.; Shazeer, N.; Parmar, N.; Uszkoreit, J.; Jones, L.; Gomez, A.N.; Kaiser, Ł.; Polosukhin, I. Attention is all you need. *Adv. Neural Inf. Process. Syst.* **2017**.
27. Battaglia, P.; Pascanu, R.; Lai, M.; Jimenez Rezende, D. Interaction networks for learning about objects, relations and physics. *Adv. Neural Inf. Process. Syst.* **2016**, *29*.
28. Koniusz, P.; Yan, F.; Gosselin, P.-H.; Mikolajczyk, K. Higher-order occurrence pooling for bags-of-words: Visual concept detection. *IEEE Trans. Pattern Anal. Mach. Intell.* **2016**, *39*, 313–326. [CrossRef] [PubMed]
29. Shih, Y.-F.; Yeh, Y.-M.; Lin, Y.-Y.; Weng, M.-F.; Lu, Y.-C.; Chuang, Y.-Y. Deep co-occurrence feature learning for visual object recognition. In Proceedings of the IEEE Conference on Computer Vision and Pattern Recognition, Honolulu, HI, USA, 21–26 July 2017; pp. 4123–4132.
30. Koniusz, P.; Zhang, H.; Porikli, F. A Deeper Look at Power Normalizations. In Proceedings of the IEEE Conference on Computer Vision and Pattern Recognition (CVPR), Salt Lake City, UT, USA, 18–23 June 2018.
31. Koniusz, P.; Tas, Y.; Porikli, F. Domain adaptation by mixture of alignments of second-or higher-order scatter tensors. In Proceedings of the IEEE Conference on Computer Vision and Pattern Recognition, Honolulu, HI, USA, 21–26 July 2017; pp. 4478–4487.
32. Božič, J.; Tabernik, D.; Skočaj, D. End-to-end training of a two-stage neural network for defect detection. In Proceedings of the 2020 25th International Conference on Pattern Recognition (ICPR), Milan, Italy, 10–15 January 2021; pp. 5619–5626.
33. Racki, D.; Tomazevic, D.; Skocaj, D. A compact convolutional neural network for textured surface anomaly detection. In Proceedings of the 2018 IEEE winter conference on applications of computer vision (WACV), Lake Tahoe, NV, USA, 12–15 March 2018; pp. 1331–1339.

34. Kim, S.; Kim, W.; Noh, Y.-K.; Park, F.C. Transfer learning for automated optical inspection. In Proceedings of the 2017 International Joint Conference on Neural Networks (IJCNN), Anchorage, AK, USA, 14–19 May 201; pp. 2517–2524.
35. Scholz-Reiter, B.; Weimer, D.; Thamer, H. Automated surface inspection of cold-formed micro-parts. *CIRP Ann.* **2012**, *61*, 531–534. [CrossRef]

applied sciences

MDPI

Article

SimMolCC: A Similarity of Automatically Detected Bio-Molecule Clusters between Fluorescent Cells

Shun Hattori [1,*], Takafumi Miki [2], Akisada Sanjo [3], Daiki Kobayashi [2] and Madoka Takahara [4]

1 Faculty of Advanced Engineering, The University of Shiga Prefecture, 2500 Hassaka-cho, Hikone 522-8533 , Japan
2 Graduate School of Medicine, Akita University, 1-1-1 Hondo, Akita 010-8543, Japan; tmiki@med.akita-u.ac.jp (T.M.); dkobayashi@med.akita-u.ac.jp (D.K.)
3 Faculty of Medicine, Akita University, Akita 010-8543, Japan; s4022556@s.akita-u.ac.jp
4 Faculty of Advanced Science and Technology, Ryukoku University, 1-5 Yokotani, Seta Oe-cho, Otsu 520-2194, Japan; takahara@rins.ryukoku.ac.jp
* Correspondence: hattori.s@e.usp.ac.jp

Abstract: In the field of studies on the "Neural Synapses" in the nervous system, its experts manually (or pseudo-automatically) detect the bio-molecule clusters (e.g., of proteins) in many TIRF (Total Internal Reflection Fluorescence) images of a fluorescent cell and analyze their static/dynamic behaviors. This paper proposes a novel method for the automatic detection of the bio-molecule clusters in a TIRF image of a fluorescent cell and conducts several experiments on its performance, e.g., mAP @ IoU (mean Average Precision @ Intersection over Union) and F1-score @ IoU, as an objective/quantitative means of evaluation. As a result, the best of the proposed methods achieved 0.695 as its mAP @ IoU = 0.5 and 0.250 as its F1-score @ IoU = 0.5 and would have to be improved, especially with respect to its recall @ IoU. But, the proposed method could automatically detect bio-molecule clusters that are not only circular and not always uniform in size, and it can output various histograms and heatmaps for novel deeper analyses of the automatically detected bio-molecule clusters, while the particles detected by the Mosaic Particle Tracker 2D/3D, which is one of the most conventional methods for experts, can be only circular and uniform in size. In addition, this paper defines and validates a novel similarity of automatically detected bio-molecule clusters between fluorescent cells, i.e., SimMolCC, and also shows some examples of SimMolCC-based applications.

Keywords: object detection; particle detection; similarity; neural synapses; computer vision

1. Introduction

In recent years, AI (Artificial Intelligence) technologies have started to become pervasive/ubiquitous in various situations of the real world (towards Society 5.0 [1], which was proposed by the Cabinet Office, Government of Japan): dialogue systems based on LLMs (Large Language Models) such as OpenAI's ChatGPT and Google's Gemini, Text-to-Image generation [2] such as Stable Diffusion and Midjourney, DX (Digital Transformation) in companies, more advanced ITSs (Intelligent Transport Systems) and automated driving, and a diverse array of AI technologies in education (such as EduTech [3]), medical care and nursing care (such as MedTech and SleepTech [4,5]), finance (such as FinTech), clothing, food, and housing [6,7], various forms of entertainment, such as sports [8] and video games [9,10], and so forth.

An Artificial Neural Network (ANN) [11], especially a Deep Neural Network (i.e., Deep Learning), which is playing a starring role in them, is a model for Machine Learning, inspired by the neuronal organization found in the biological neural networks in animal brains. An ANN is composed of connected units called artificial neurons, which loosely model the natural neurons in a brain and receive signals from connected artificial neurons, process the signals, and send signals to other connected artificial neurons. Artificial neurons are connected by edges, which model the "Neural Synapses" in a brain. Therefore, more advancement in Brain Science contributes to more advancement in ANNs and Brain Computing.

In such a field of studies as the "Neural Synapses" in the nervous system, its experts observe and manually (or pseudo-automatically) detect bio-molecule clusters (e.g., of proteins) in many TIRF (Total Internal Reflection Fluorescence) images of a fluorescent cell and analyze their static/dynamic behaviors. Most of the conventional methods for "Bright Spot Analysis" and "Particle Tracking" fit the point spread function of not multiple but a single fluorescent particle to a 2D Gaussian function and apply template matching to an input image [12], e.g., the Mosaic Particle Tracker 2D/3D [13].

As shown in Figure 1, information transmission in the nervous system occurs at the Neural Synapses, which conjugate neuron cells:

- In the presynaptic terminal of a neuron cell, which transmits information, synaptic vesicles are recruited to the release sites at the active zone and release message-carrying chemicals rapidly upon Ca^{2+} influx outside the neuron cell towards its corresponding postsynaptic site. Note that the docked vesicles at the release sites are considered to be the vesicles within the so-called RRP (Readily Releasable Pool) [14];
- In the postsynaptic site of the corresponding neuron cell, which receives information, the released message-carrying chemicals act on postsynaptic receptors and evoke postsynaptic responses.

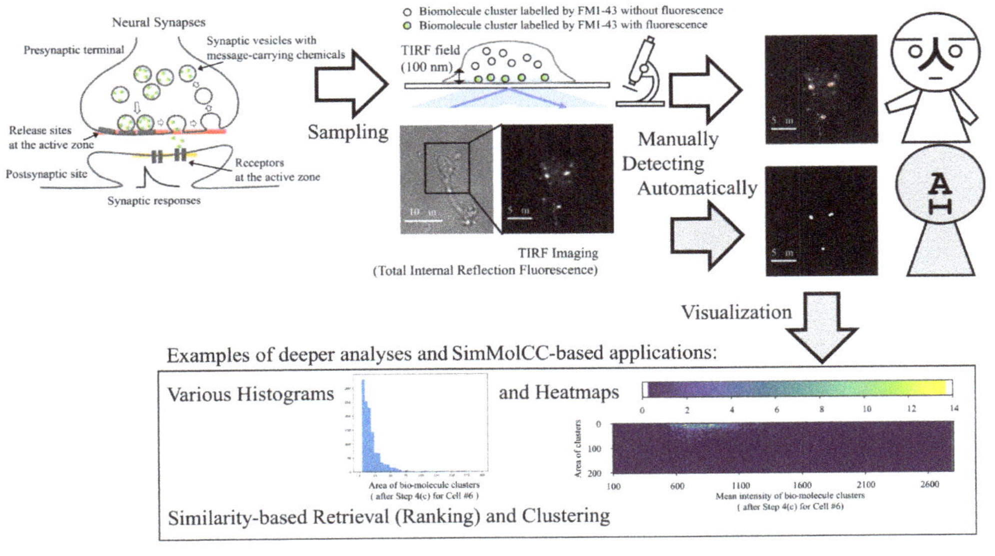

Figure 1. Direct imaging bio-molecule clusters in a fluorescent cell, e.g., in the presynaptic terminal of a neuron cell, using TIRF (Total Internal Reflection Fluorescence) microscopy and detecting them manually by human experts or automatically by the proposed method.

TIRF microscopy is explained as follows in Chapter 13 [14] of the book whose title is "Exocytosis from molecules to cells", published by IOP Publishing:

> Zenisek et al. [15] pioneered the imaging of single-vesicle dynamics in dissociated goldfish retinal bipolar cells using total internal reflection fluorescence (TIRF) microscopy. TIRF microscopy has very good resolution in the z-axis (50–100 nm), which provides detailed vesicle dynamics near the plasma membrane. On the other hand, the resolutions of the x- and y-axes are diffraction-limited, which means that single vesicles (30–50 nm in diameter) appear as single dots. Sparse labeling of synaptic vesicles with FM1-43 (FM1-43 emits fluorescence

when excited by TIRF microscopy's Evanescent field.) allows one to look at the dynamics of synaptic vesicles before and during fusion. One can see the fusion of synaptic vesicles, which accompanies the loss of dyes in the center and a transient increase in the surrounding fluorescence, reflecting the diffusion of dyes along the plasma membrane.

Moreover, an example of observation (especially direct imaging) by TIRF microscopy of a target phenomenon, e.g., rapid tethering, of synaptic vesicles accompanying exocytosis at a fast central synapse is also shown as follows:

> Miki et al. [16] applied TIRF microscopy to cerebellar mossy fiber terminals. They found that the RRP (Readily Releasable Pool) corresponds to those vesicles already resident and ready for fusion upon Ca^{2+} influx. Following depletion of the RRP, vesicles which are within the TIRF field (100 nm) are fused in response to sustained depolarization or a train of action potentials, suggesting that newly replenished vesicles are already close to the membrane. At the same time, vesicles are recruited to the TIRF field more rapidly than they are to the calyx synapse. In addition, newly tethered vesicles can be fused with maturation times of several hundreds of milliseconds, which is much faster than that of the calyx of Held. Therefore, cerebellar mossy fiber terminals have more efficient vesicle recruitment and priming processes than those of the calyx of Held synapse.

For more advances in Brain Science with more advances in ANNs and Brain Computing, in the nervous system, the following features, i.e., properties and static/dynamic behaviors, of bio-molecule clusters in many TIRF images of a fluorescent cell need to be observed and analyzed from various directions.

- Properties: each bio-molecule cluster's size (e.g., the width and height of its detected Bounding Box), segmented area, shape (e.g., circle/spot-like, narrow, or odd-looking), 2D/3D position (x-, y-, and z-axes), fluorescence intensity, etc.
- Static/dynamic behaviors: each bio-molecule cluster's change in state, tethering at the active zone, releasing message-carrying chemicals, vanishing from the active zone, moving in a cell, receiving and responding to message-carrying chemicals, and fusing with a membrane or with the other bio-molecule cluster(s), etc.

In addition, there are various methods of analysis, as indicated: temporal analysis, spatial analysis, spatio-temporal analysis, state analysis, similarity analysis (e.g., retrieving similar cells with similar features, or clustering cells based on a similarity between cells), network/community analysis, etc.

However, huge costs, e.g., a long time for manually detecting the bio-molecule clusters in many TIRF images of a fluorescent cell as well as a large sum of money for making them by TIRF microscopy, have been hindering speeding up the advancements in Brain Science, and there are other problems including biased detection and missing some of them. Therefore, as shown in Figure 1, this paper proposes a novel method for the automatic detection of the bio-molecule clusters in a TIRF image of a fluorescent cell to reduce the manual costs and solve the manual problems, and several experiments have been conducted on its performance, e.g., mAP @ IoU (mean Average Precision @ Intersection over Union) and F1-score @ IoU, as an objective/quantitative means of evaluation. The proposed method can automatically detect bio-molecule clusters that are not only circular and not always uniform in size, and it can output various histograms and heatmaps for novel deeper analyses of the automatically detected bio-molecule clusters, while the particles detected by the Mosaic Particle Tracker 2D/3D [13] can be only circular and uniform in size. In addition, this paper defines and validates a novel similarity of automatically detected bio-molecule clusters between fluorescent cells, i.e., SimMolCC, and also shows some examples of SimMolCC-based applications.

The remainder of this paper is organized as follows. Section 2 introduces two kinds of related studies and compares them with this paper. Section 3 describes the novel method in detail for the automatic detection of the bio-molecule clusters in a TIRF image of a fluorescent cell. Section 4 defines a novel similarity of automatically detected bio-molecule clusters between fluorescent cells, i.e., SimMolCC. Section 5 shows several experimental results to validate the proposed method. Finally, Section 6 concludes this paper.

2. Related Work

2.1. Object Detection on Cells

In such a broad field of studies on general-purpose/specific "Computer Vision", many object detection and segmentation techniques, e.g., YOLO (You Only Look Once) [17], and their practical applications have been proposed [2,18–22]: automatic driving/traffic [23–26], maritime [27,28], aerial [29], remote sensing [30], agriculture [31,32], and power line infrastructure [33].

Meanwhile, many object detection and segmentation techniques that are not general-purpose but specific to (the region of) cells have also been proposed: from classical techniques [34] based on conditional opening and closing [35], Laplace edge features and SVM (Support Vector Machine) [36], HOG (Histogram of Oriented Gradients) features and SVM [37], SIFT (Scale-Invariant Feature Transform) features, Random Forests, and Hierarchical Clustering [38], or other features [39], to Deep Learning techniques [40,41] such as cellpose [42], Residual U-Net [43,44], which combines U-Net [45] and Residual-Net [46], and R2U-Net [47], which is a Recurrent Residual convolutional neural network based on U-Net.

However, few object detection and segmentation techniques to detect the micro-objects in a cell, e.g., a nucleus in a cell [36,48–51], and melanin [52–54] in a microscopic image of the stratum corneum for skin diagnosis, have been proposed, as shown in Table 1. This paper proposes novel methods to detect the nano-objects, e.g., bio-molecule clusters (of proteins) in a TIRF (Total Internal Reflection Fluorescence) image of a cell for neuroscience, and to analyze their size and fluorescence intensity, e.g., as various histograms (size/area $\rightarrow$ frequency, or intensity $\rightarrow$ frequency) and heatmaps (size/area $\times$ intensity $\rightarrow$ frequency).

Table 1. Comparison between methods to detect micro-objects in a cell.

	[51]	[52]	This Paper
Targets	Cell Nuclei	FM (Fontana-Masson) Stained Melanin	Bio-Molecule Clusters [*1]
	in Blue/red-stained Buccal Cells for Liquid Cytology 1280 × 1024 [pixels] 21? nm per pixel [*2] TIFF (Full? Color)	in Face Epidermal Corneocyte 736 × 440 [pixels] 272 nm per pixel BMP (24-bit RGB)	in a Fluorescent Cell 512 × 512 [pixels] 65 nm per pixel TIFF (16-bit Grayscale)
Input image	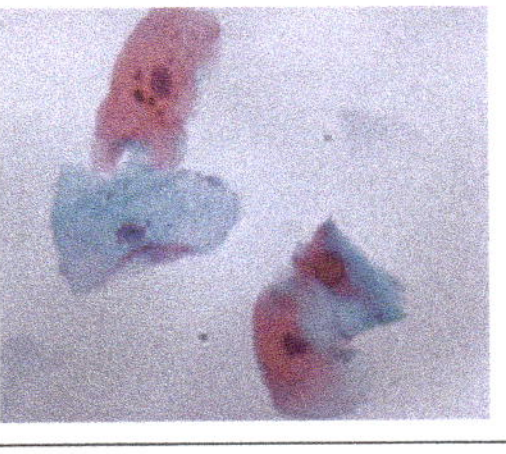	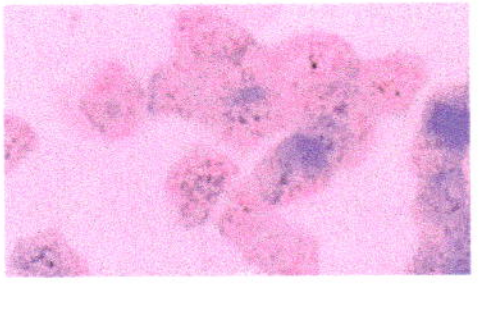	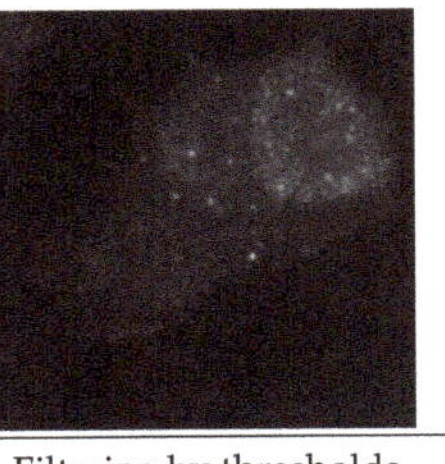
Main techs to detect micro-objects in a cell	Sliding Window Method Mask-RCNN	Template matching	Filtering by thresholds Edge extraction

Table 1. *Cont.*

	[51]	[52]	This Paper
Output image	 red: cell nuclei with a probability over 0.90 yellow: cell nuclei with a probability over 0.50	 black: background gray: laminated Corneocyte white: Corneocyte pink: Melanin spots	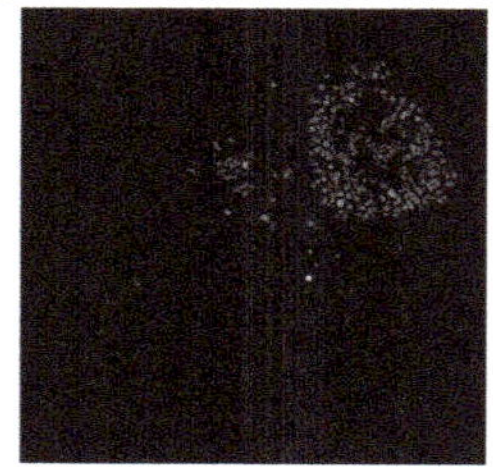 black: not bio-molecule clusters gray: bio-molecule clusters with a fluorescence intensity

*1 The size of 1 target bio-molecule is about 10 nm, observed as 200–300 nm (2D Gaussian, σ = 120–130 nm) because the TIRF's resolutions of x- and y-axes are diffraction-limited. Therefore, it is applied to Step 4(c) of the proposed method that a target bio-molecule cluster would occupy at least 5 pixels in an input TIRF image.
*2 Maybe 21 nm per pixel = $271.7 \cdot 736/300 \cdot 40/1280$, which is not clearly specified in [51] but is estimated from 40× in [51], while 300× in [52].

2.2. Similarity on Cells

In the field of studies on general-purpose "Image Recognition" and "Content-Based Image Retrieval (CBIR)", various graphic similarities between images based on their bag of features (i.e., visual words) have been defined [55]. The image features are divided into two kinds: global features, e.g., a color histogram [56], which are extracted by globally describing the features of an image, while the local features, e.g., SIFT (Scale-Invariant Feature Transform) [57], SURF (Speeded Up Robust Features) [58], HOG (Histogram of Oriented Gradients) [37], and LBP (Local Binary Pattern) [56], are extracted by detecting the points of the local features in an image and locally describing the feature for each point. In recent years, image features based on DNNs (Deep Neural Networks) have also been proposed. DELG [59] unifies deep local and global features for Google Landmark Recognition.

Meanwhile, a few similarities between those images that are not general-purpose but specific to cells have been defined. CellSim [60] has been developed as a software of bioinformatics for researchers to calculate the similarity between different cells by the semantic similarity algorithm [61] based on the cell ontology network and cell-specific regulation network in over 2000 different human cell types, e.g.,

> Auditory Epithelial Cell; Blood Progenitor Cell; Connective Tissue Cell; Dendritic Cell; Embryo Cell; Epithelial Cell; Epithelial Stem Cell and Muscle Myoblast; Germ Cell; Germ Cell and Spore; Hematopoietic Cell; Keratinocyte Cell; Kidney Cell (part); Kidney Epithelial; Lymphocyte; Macrophage; Marrow Cell; Microfold Cell; Muscle Cell; Myoepithelial Cell; Neurecto-epithelial Cell; Neurogliocyte; Neuron; Neuron Cell; Osteoblast Mesenchymal Stem Cell; Pigment Cell; Secreting Cell; Sensory Epithelial Cell; Somatic Stem Cell; Step Cell (mixed); and Vessel Endothelial,

from FANTOM Ontology [62] and provides the sharing regulation networks of part cells. CellSim can also predict cell types by inputting a list of genes as a query, including more than 250 human normal-tissue-specific cell types and 130 cancer cell types, and provide the prediction results in both tables and spider charts, which can be preserved easily and freely.

The proposed similarity, SimMolCC, in this paper is a graphic similarity between instances of cells by automatically detecting the bio-molecule clusters in an image of a fluorescent cell and describing its global features based on their size/area, fluorescence intensity, ratio of width to height of Bounding Box, and ratio of area to Bounding Box, while CellSim is a semantic similarity between types (i.e., classes in the context of "Object-Orientation"; categories in the context of "Image Categorization") of cells.

For your information, two kinds of SimCells are not related to this paper: one [63], developed at the Tokyo Institute of Technology, is a processor simulator for multi-core architecture research, and the other [64], developed at the University of Oxford, is a platform for human health – cells made simple.

3. Automatic Detection of Bio-Molecule Clusters in a Fluorescent Cell Image

This section describes in detail novel methods for automatic detection of bio-molecule clusters in a TIRF image of a fluorescent cell.

3.1. Overview

Figure 2 provides an overview of the proposed methods, which have the following input and outputs (as shown in Figure 3), and the following five steps (with fourteen sub-steps):

Input is a TIRF image (.tif, unsigned 16-bit grayscale, 512 × 512 [pixels]) of a fluorescent cell.

Outputs are bio-molecule clusters in a fluorescent cell, and also their size/area, fluorescence intensity, ratio of area to Bounding Box, and ratio of width to height of Bounding Box. In addition, various histograms (size/area → frequency, etc.) and heatmaps (size/area × intensity → frequency, etc.) can be outputted.

Step 1. Segmenting the target cell in an input TIRF image (described in Section 3.2).

- **Step 1(a).** Filtering out pixels outside the target cell by an automatically calculated threshold θ_{step1a}.
- **Step 1(b).** Averaging (i.e., filtering out some sort of noise).

Step 2. Segmenting and dividing the regions of the target bio-molecule clusters in an input TIRF image of a fluorescent cell (described in Section 3.4).

- **Step 2(a).** Filtering out pixels that seem not to be candidates for the target bio-molecule clusters of a fluorescent cell by an automatically calculated threshold θ''_{step1a} at Step 1(a).
- **Step 2(b).** Laplacian edge extraction with the size of kernel, `kernel_size` $\in [1, 13]$.
- **Step 2(c).** Dividing all the regions of the target bio-molecule clusters into each region of bio-molecule cluster.

Step 3. Clustering and assigning the regions of the target bio-molecule clusters in an input TIRF image of a fluorescent cell with their ID (described in Section 3.5).

- **Step 3(a).** Canny edge extraction for the target cell's edges and the target bio-molecule clusters' edges by applying Otsu method [65,66].
- **Step 3(b).** Filtering Canny edges out from the target bio-molecule clusters to make them independent.
- **Step 3(c).** Clustering and assigning the target bio-molecule clusters with their ID (Identification Data).
- **Step 3(d).** Integrating Canny edges filtered out at Step 3(b) back into one of the target bio-molecule clusters.

Step 4. Filtering bio-molecule clusters (described in Section 3.6).

- **Step 4(a).** Filtering out bio-molecule clusters that do not touch any Canny edges (i.e., any outline of candidates for bio-molecule clusters) in the target cell in an input TIRF image.
- **Step 4(b).** Filtering out bio-molecule clusters that touch the Canny edge (i.e., the outline) of the target cell in an input TIRF image.
- **Step 4(c).** Filtering out bio-molecule clusters whose size/area is less than 5 pixels.
- **Step 4(d).** Filtering bio-molecule clusters based on their fluorescence intensity.

Step 5. Calculating the size/area, fluorescence intensity, ratio of area to Bounding Box, and ratio of width to height of Bounding Box of each automatically detected bio-molecule cluster, and also creating various histograms and heatmaps of automatically detected bio-molecule clusters as visualization.

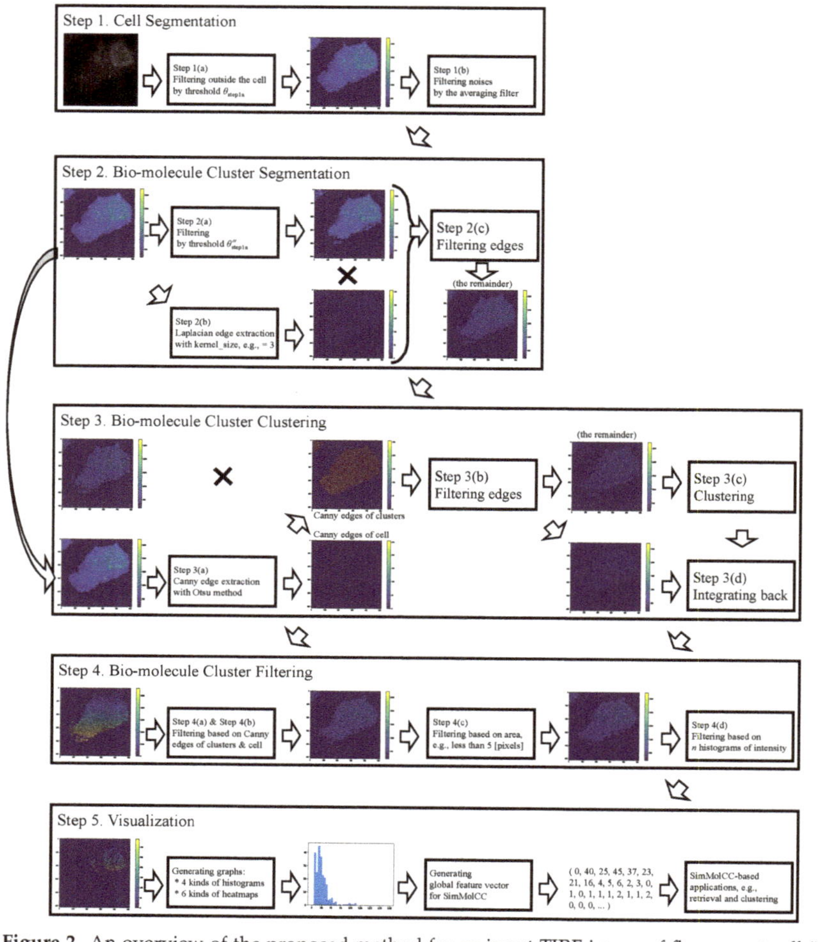

Figure 2. An overview of the proposed method for an input TIRF image of fluorescent cell #6 to automatically detect its bio-molecule clusters by Steps 1 to 4 and to output histograms, heatmaps, and its global feature vector for SimMolCC by Step 5.

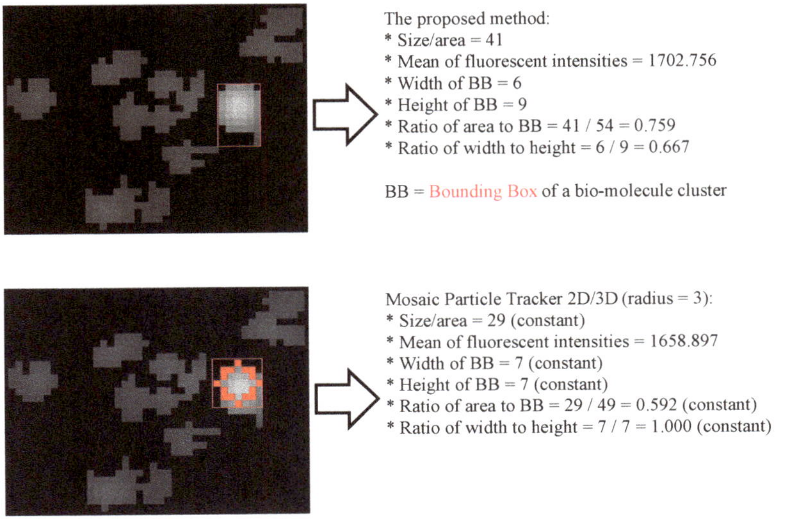

Figure 3. Comparison of a bio-molecule cluster's features between the proposed method and Mosaic Particle Tracker 2D/3D [13] for an input TIRF image of fluorescent cell #6.

3.2. *Step 1—Cell Segmentation*

Step 1 segments the target cell in an input TIRF image by the following two sub-steps as precisely as possible as shown in Figure 4:

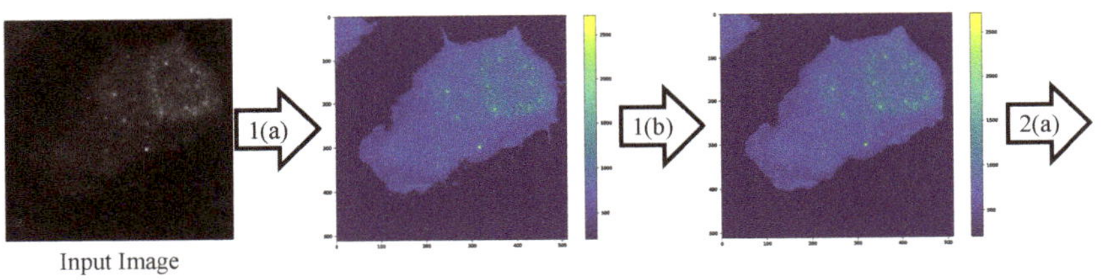

Figure 4. Step 1 has two sub-steps, Step 1(a) and Step 1(b), to segment the target cell in an input TIRF image (fluorescent cell #6) as precisely as possible.

Step 1(a). First, the histogram of fluorescence intensity of each pixel $\in 512 \times 512$ [pixels] in an input TIRF image of a fluorescent cell is calculated as shown in Figure 5, where the number of bins, `bins`, is set based on the following Sturges' rule [67]:

$$\text{Sturges' optimal number of bins} = \lceil \log_2 N + 1 \rceil \quad (1)$$

where N means the number of samples for the histogram, e.g., $N = 512 \cdot 512$, at the initial Step 1, and the symbol $\lceil x \rceil$ means "ceiling", i.e., round the answer x up to the nearest integer. As a result, Sturges' optimal number of bins is always calculated as 20 at Step 1(a).

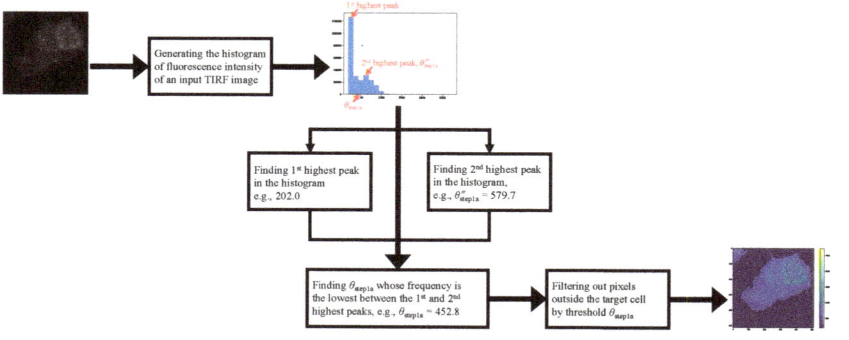

Figure 5. A flowchart of Step 1(a) with the histogram of fluorescence intensity of each pixel $\in 512 \times 512$ [pixels] in an input TIRF image of fluorescent cell #6.

Next, peaks are found in the histogram by `find_peaks()` of signal processing of SciPy [68], and the local minimum between the 1st- and 2nd-highest peaks is also found. For example, in the histogram as shown in Figure 5,

- the 1st-highest peak's fluorescence intensity is 202.0.
- the 2nd-highest peak's fluorescence intensity $\theta''_{\mathrm{step1a}}$ is 579.7.
- the fluorescence intensity θ_{step1a} whose frequency is the lowest between the 1st- and 2nd-highest peaks is 452.8.

Finally, any pixel of the input TIRF image of a fluorescent cell whose fluorescence intensity is lower than or equal to the above-calculated fluorescence intensity θ_{step1a} of the local minimum between the 1st- and 2nd-highest peaks is filtered out as shown in Figure 5.

Step 1(b). The filtered TIRF image of a fluorescent cell by the fluorescence intensity θ_{step1a} is averaged by the averaging filter, whose size of kernel is set to 19×19, and any pixel of the filtered TIRF image of a fluorescent cell whose averaged value is lower than or equal to $\frac{14 \cdot 14}{19 \cdot 19} \risingdotseq 0.543$ is filtered out as shown in Figure 4 because some sort of noise, e.g., salt-and-pepper noise, has to be filtered out.

3.3. Step 2—Bio-Molecule Cluster Segmentation

Step 2 segments and divides the regions of the target bio-molecule clusters in an input TIRF image of a fluorescent cell by the following three sub-steps as precisely as possible as shown in Figure 6:

Step 2(a). Any pixel of the filtered TIRF image of a fluorescent cell after Step 1(b) whose fluorescence intensity is lower than or equal to the 2nd-highest peak's fluorescence intensity $\theta''_{\mathrm{step1a}}$ at Step 1(a) is filtered out. The remainder seems to include not-independent candidates for the target bio-molecule clusters of the fluorescent cell.

Step 2(b). Laplacian edges are extracted from the filtered TIRF image of a fluorescent cell after Step 1(b) by OpenCV's Laplacian operator [69], `cv2.Laplacian()`, which has the size of kernel to be optimized, `kernel_size` $\in [1, 13]$, in this paper.

Step 2(c). Laplacian edges are filtered out from the filtered TIRF image of a fluorescent cell after Step 2(a) in order to divide all the regions of candidates for the target bio-molecule clusters of the fluorescent cell into each region of bio-molecule cluster.

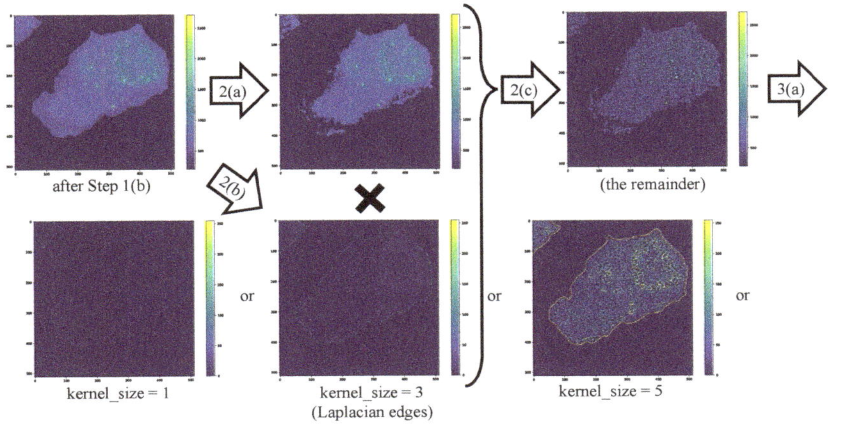

Figure 6. Step 2 has three sub-steps, Step 2(a), Step 2(b), and Step 2(c), to segment the regions of bio-molecule clusters in an input TIRF image of fluorescent cell #6 as precisely as possible.

3.4. Step 3—Bio-Molecule Cluster Clustering

Step 3 divides the regions of the target bio-molecule clusters in an input TIRF image of a fluorescent cell into each region of bio-molecule cluster by the following four sub-steps as precisely as possible as shown in Figure 7:

Step 3(a). First, Canny edges are extracted from the filtered TIRF image of a fluorescent cell after Step 1(b) by OpenCV's Canny [70], `cv2.Canny()`, which has the first and second thresholds to be optimized. This paper automatically optimizes the two thresholds by applying Otsu method [65,66].
Next, Canny edges are divided into the target cell's ones or the target bio-molecule clusters' ones depending on whether or not they touch any pixel outside the target cell, which has already been filtered out and thus whose intensity has already been set to "0 (zero)."

Step 3(b). Canny edges are filtered out from the filtered TIRF image of a fluorescent cell after Step 2(c) in order to divide all the regions of candidates for the target bio-molecule clusters of the fluorescent cell into each region of bio-molecule cluster. The remainder seems to include independent candidates for the target bio-molecule clusters of the fluorescent cell.

Step 3(c). The regions of the target bio-molecule clusters in the filtered TIRF image of a fluorescent cell after Step 3(b) have "Clustering" applied. As a result, each bio-molecule cluster becomes independent and is assigned the sequential ID (Identification Data); e.g., the number of bio-molecule clusters is calculated as 5507 in Figure 7.

Step 3(d). Canny edges filtered out at Step 3(b) are integrated back into one of the target bio-molecule clusters in the filtered TIRF image of a fluorescent cell after Step 3(c). Note that the number of bio-molecule clusters at Step 3(c) and also at Step 3(d), e.g., 5507, seems to be too many. Therefore, the following Step 4 is required.

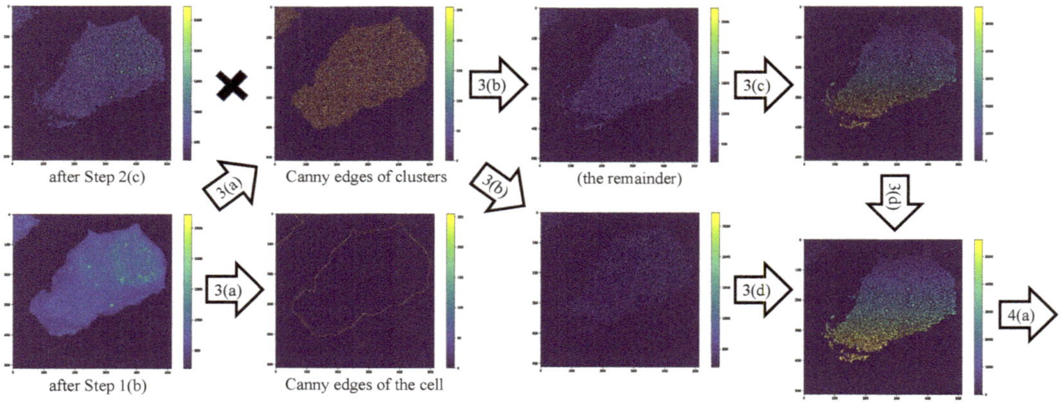

Figure 7. Step 3 has four sub-steps, Step 3(a), Step 3(b), Step 3(c), and Step 3(d), to divide the regions of bio-molecule clusters in an input TIRF image of fluorescent cell #6 as precisely as possible, and finally each bio-molecule cluster is independent and assigned the sequential ID.

3.5. Step 4—Bio-Molecule Cluster Filtering

Step 4 filters bio-molecule clusters by the following four heuristic rules:

Step 4(a). "Correct bio-molecule clusters have to have their edge (i.e., outline) in the target cell." Therefore, Step 4 filters out bio-molecule clusters that do not touch any Canny edges (i.e., any outline of candidates for bio-molecule clusters) in the target cell in an input TIRF image.

Step 4(b). "Correct bio-molecule clusters have not to exist in protrusions near the edge (i.e., outline) of the target cell." Therefore, Step 4 filters out bio-molecule clusters that touch the Canny edge (i.e., the outline) of the target cell in an input TIRF image.

Step 4(c). "The size of 1 correct bio-molecule is about 10 nm, observed as 200–300 nm (2D Gaussian, $\sigma = 120$–130 nm)" because the TIRF's resolutions of x- and y-axes are diffraction-limited. Therefore, Step 4 filters out bio-molecule clusters whose area is less than 5 [pixels] in an input TIRF image of a fluorescent cell.

Step 4(d). "Correct bio-molecule clusters have to have unusually higher fluorescence intensity in the target cell." First, the n kinds of sampled histograms of fluorescence intensity of each pixel that has not yet been filtered out and thus whose value has not yet been "0 (zero)" in the filtered TIRF image of a fluorescent cell after Step 4(c) are calculated, where the number of bins, `bins`, is set based on the Sturges' optimal number of bins [67] from $+0$ to $+(n-1)$, as shown in Figure 8. Note that the number n of sampled histograms is set to 5 in this paper.

Next, the threshold to filter out bio-molecule clusters that do not have unusually higher fluorescence intensity in the target cell is automatically searched by either of the following two kinds of ways:

1st: The threshold flagged as "1st" is set to be the average of the n fluorescence intensities of the bin that first violates "Monotone Decreasing" in each sampled histogram.

3rd: The threshold flagged as "3rd" is set to be the average of the n fluorescence intensities of the bin that violates "The difference of frequency (between the bin and the bin followed by it) is not 3rd compared with its pre-difference and its post-difference" in each sampled histogram.

Note that the number of bio-molecule clusters at Step 4(d) (`kernel_size` = 3 flagged as "3rd") is calculated as 237 in Figure 8.

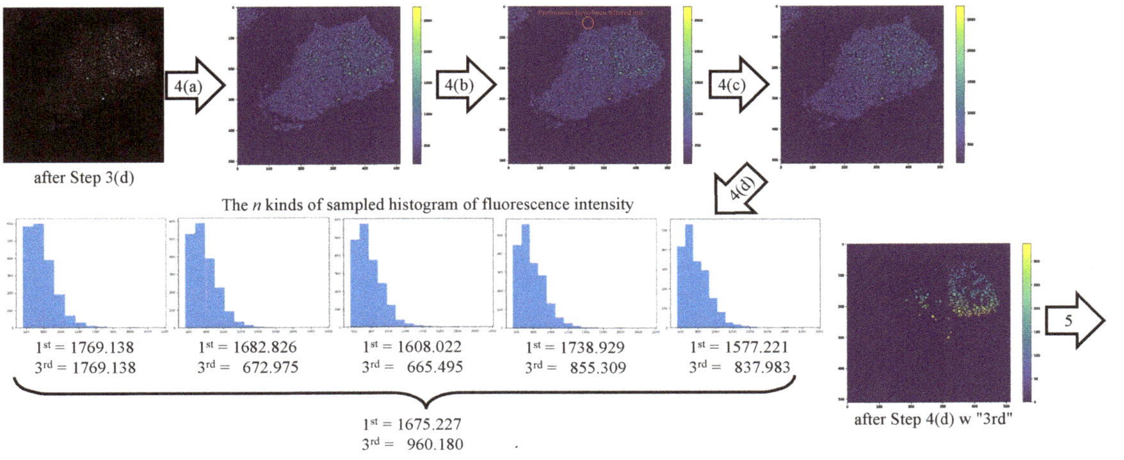

Figure 8. Step 4 has four sub-steps, Step 4(a), Step 4(b), Step 4(c), and Step 4(d), to filter bio-molecule clusters by four kinds of heuristic rules.

3.6. Step 5—Visualization

Step 5 can calculate four kinds of features such as the size/area, fluorescence intensity, ratio of area to Bounding Box, and ratio of width to height of Bounding Box of each automatically detected bio-molecule cluster in the target cell of an input TIRF image at each above-mentioned step, and also create various histograms and heatmaps of automatically detected bio-molecule clusters as visualization. For example, Figure 9 shows the four kinds of histograms of the size/area, fluorescence intensity, ratio of area to Bounding Box, and ratio of width to height of Bounding Box of each of the 237 automatically detected bio-molecule clusters in an input TIRF image of a fluorescent cell #6 at Step 4(d) (`kernel_size` = 3 flagged as "3rd"), and Figure 10 shows the six kinds of heatmaps between four kinds of features such as the size/area, fluorescence intensity, ratio of area to Bounding Box, and ratio of width to height of Bounding Box of each of the 237 automatically detected bio-molecule clusters in an input TIRF image of a fluorescent cell #6 at Step 4(d) with the size of kernel, `kernel_size` = 3, for OpenCV's Laplacian operator and flagged as "3rd." These figures could help experts to conduct deeper analyses of bio-molecule clusters in a TIRF image of a fluorescent cell.

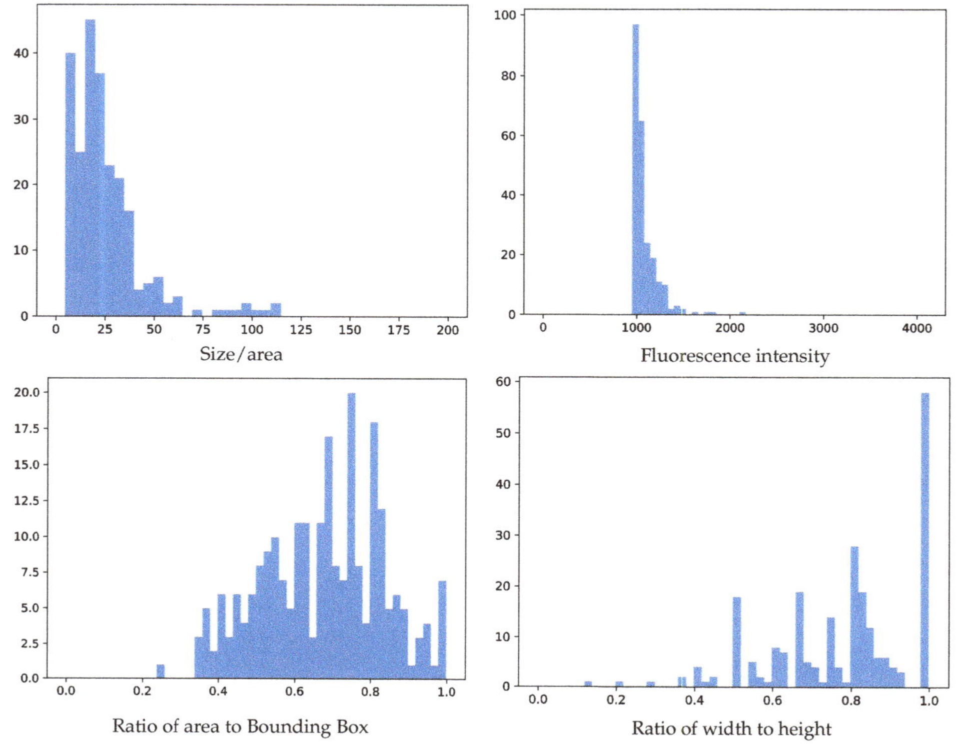

Figure 9. The four kinds of histograms of the size/area, fluorescence intensity, ratio of area to Bounding Box, and ratio of width to height of Bounding Box of each of the 237 automatically detected bio-molecule clusters in an input TIRF image of fluorescent cell #6 at Step 4(d) with the size of kernel, `kernel_size` = 3, for OpenCV's Laplacian operator and flagged as "3rd".

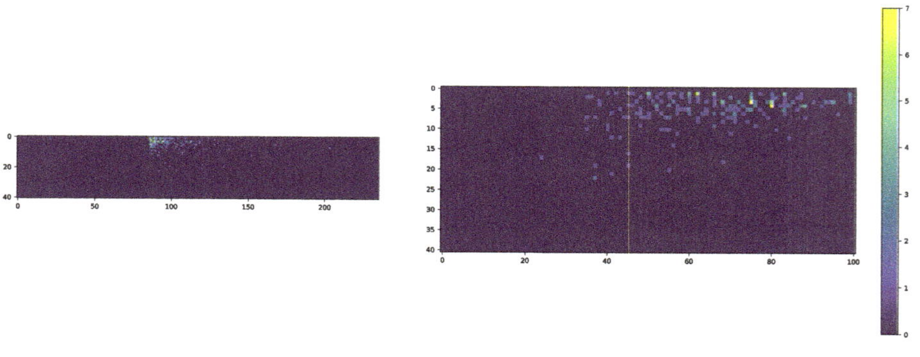

Area $y\times$ Intensity x Area $y\times$ Ratio of area to BB x

Figure 10. *Cont.*

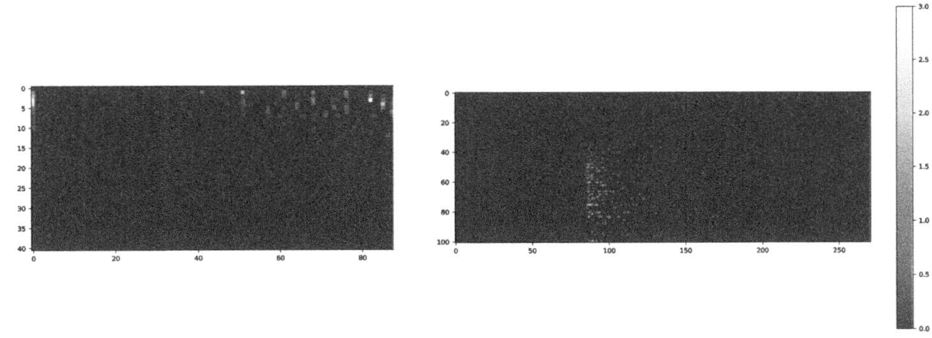

Area $y\times$ Ratio of width to height x | Ratio of area to BB $\times$ Intensity x

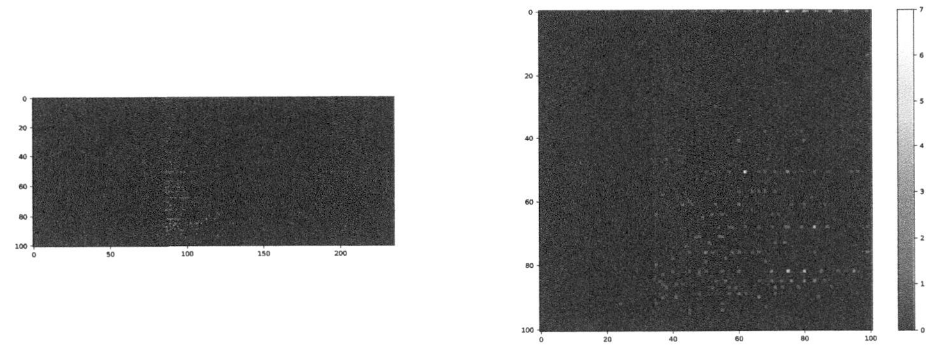

Ratio of width to height $y\times$ Intensity x | Ratio of width to height $y\times$ Ratio of area to BB

Figure 10. The six kinds of heatmaps between four kinds of features, such as the area, fluorescence intensity, ratio of area to Bounding Box, and ratio of width to height of Bounding Box of each of the 237 automatically detected bio-molecule clusters in an input TIRF image of fluorescent cell #6.

4. SimMolCC: Similarity of Automatically Detected Bio-Molecule Clusters between Fluorescent Cells

This section defines a novel similarity of automatically detected bio-molecule clusters between fluorescent cell images, i.e., SimMolCC, as follows:

$$\text{SimMolCC}(img_1, img_2) \;:=\; \text{cosine-similarity}(\vec{v_1}, \vec{v_2}) \tag{2}$$

$$:= \frac{\vec{v_1}\cdot\vec{v_2}}{||\vec{v_1}||\cdot||\vec{v_2}||} \tag{3}$$

where $\vec{v_1}$, $\vec{v_2}$ mean each global feature vector extracted from an input TIRF image of a fluorescent cell.

In the following Experiment II, the four kinds of histograms will be adopted as the global feature vector $\vec{v_i}$ of each input TIRF image of a fluorescent cell. For example,

- `area`: the histogram of size/area of Figure 9, where `range` $=(0,200)$ and `bins` $=40$, is converted to the 40-dimensional global feature vector $\vec{v_i}$ of an input TIRF image of a fluorescent cell #6, $(0,40,25,45,37,23,21,16,4,5,6,2,3,0,1,0,1,1,1,2,1,1,2,0,0,0,...)$;
- `intensity`: the histogram of mean fluorescence intensity of Figure 9, where `range` $=(0,4096)$ and `bins` $=64$, is converted to the 64-dimensional global feature vector $\vec{v_i}$ of an input TIRF image of a fluorescent cell #6, (0, ..., 0, 97, 65, 24, 19, 11, 10, 2, 3, 2, 0, 1, 0, 1, 1, 0, 0, 0, 0, 1, 0, 0, 0, ...);

- `ratio_area_BB`: the histogram of ratio of area to Bounding Box of Figure 9, where `range` $= (0.0, 1.0)$ and `bins` $= 50$, is converted to the 50-dimensional global feature vector $\vec{v}_i$ of an input TIRF image of a fluorescent cell #6, (0, ..., 0, 1, 0, 0, 0, 0, 3, 5, 2, 6, 3, 6, 4, 6, 8, 9, 10, 7, 5, 11, 11, 3, 11, 17, 8, 7, 20, 8, 4, 18, 12, 5, 6, 5, 1, 3, 4, 1, 7);
- `ratio_width_height`: the histogram of ratio of area to Bounding Box of Figure 9, where `range` $= (0.0, 1.0)$ and `bins` $= 50$, is converted to the 50-dimensional global feature vector $\vec{v}_i$ of an input TIRF image of a fluorescent cell #6, (0, 0, 0, 0, 0, 0, 1, 0, 0, 0, 1, 0, 0, 0, 1, 0, 0, 0, 2, 0, 4, 1, 2, 0, 0, 18, 0, 5, 2, 1, 8, 7, 0, 19, 5, 4, 1, 14, 4, 1, 28, 19, 12, 6, 6, 4, 3, 0, 0, 58).

Note that the six kinds of heatmaps and various hybrids with some of the four kinds of histograms and/or some of the six kinds of heatmaps can also be adopted as the global feature vector $\vec{v}_i$ of each input TIRF image of a fluorescent cell, and note that frequencies of a global feature vector can be converted by the $log_2()$ function, like TF–IDF (Term Frequency–Inverse Document Frequency).

5. Experiments

This section shows the experimental results to validate the two kinds of proposed methods in this paper:

Experiment I on automatic detection of bio-molecule clusters in a fluorescent cell image (as described in Section 3).

Experiment II on SimMolCC, a similarity of automatically detected bio-molecule clusters between fluorescent cell images (as described in Section 4).

5.1. Datasets

As shown in Figure 11, the dataset, Dataset I, for Experiment I on automatically detected bio-molecule clusters in a fluorescent cell image, has 15 sets of the following data:

1. A raw fluorescent cell movie (.tif) consisting of 100 frames (unsigned 16-bit grayscale, 512 × 512 [pixels]).
2. An averaged fluorescent cell image (.tif, unsigned 16-bit grayscale, 512 × 512 [pixels]) by Fiji's Z Projection [71] with "Average Intensity" as the projection type. Note that it is used as an input image to the proposed method for automatic detection of bio-molecule clusters in a fluorescent cell image.
3. An averaged fluorescent cell image (.tif, unsigned 24-bit RGB, 512 × 512 [pixels]) with its particles detected by the Mosaic Particle Tracker 2D/3D [13] with the parameters, `radius` $= 3$ (default), `Cutoff` $= 0.001$ (default), and `Per/Abs` (absolute is unchecked and not used. The parameter `Per`, which means percentile to determine which intense (bright) pixels are accepted as particles, was set to 0.50 (default) or 0.80 resultantly.) optimized manually by the 3rd author. Note that it tends to include noisy particles, e.g., particles outside the target cell and particles in protrusions near the edges of the target cell, and has not yet been able to be adopted as a ground truth for the proposed method for automatic detection of bio-molecule clusters in a fluorescent cell image, and also note that its particles detected by the Mosaic Particle Tracker 2D/3D [13] can be only circular and uniform in size, while the proposed method could automatically detect bio-molecule clusters that are not only circular and not always uniform in size.
4. An averaged fluorescent cell image (.tif, unsigned 24-bit RGB, 512 x 512 [pixels]) with its particles filtered manually by the 1st author and checked by the 2nd author. Note that it filtered noisy particles out, e.g., particles outside the target cell and particles in protrusions near the edges of the target cell, as precisely and exhaustively as possible, and has been adopted as a ground truth for the proposed method for automatic detection of bio-molecule clusters in a fluorescent cell image.

Note that plasma membranes of HEK293 cells attached to a coverslip were stained with wheat germ agglutinin lectin conjugated fluorescent dye (CF488 WGA Dye, biotium), and images of the membranes attached to the coverslip were acquired by TIRF (Total Internal Reflection Fluorescence) microscopy.

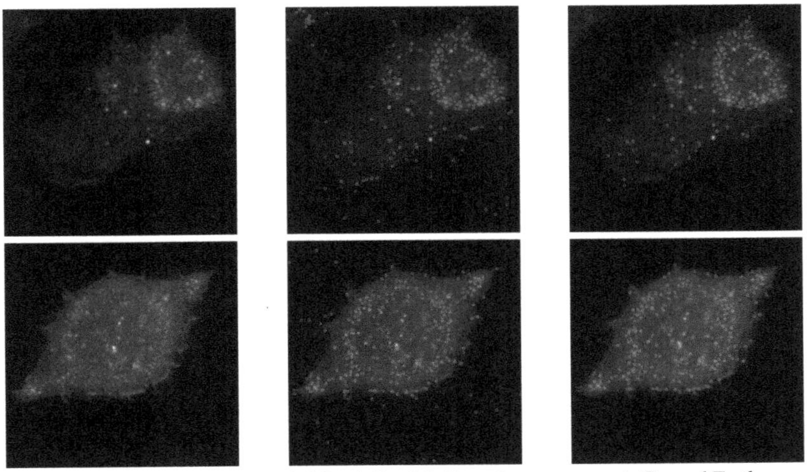

2. Input Image 3. Mosaic Particle Detection 4. Ground Truth

Figure 11. The averaged image (i.e., an input image for the proposed method), the averaged image with its particles detected by the Mosaic Particle Tracker 2D/3D [13], and the averaged image with its particles filtered manually (i.e., a ground truth for the proposed method) of an input movie (fluorescent cell #6 or #14).

The dataset, Dataset II, for Experiment II on SimMolCC, a similarity of automatically detected bio-molecule clusters between fluorescent cell images, has 105 (= ${}_{15}C_2$) similarities on bio-molecule clusters between the above-mentioned 15 averaged fluorescent cell images and 15 similarities on bio-molecule clusters between each of the 15 averaged fluorescent cell images and itself; i.e., the latter 15 similarities should be recognized as 100% (perfectly matched) by human subjects. Each similarity of bio-molecule clusters between two averaged fluorescent cell images is 11-grade-evaluated by two of three human subjects: one expert and one candidate for an expert on Cell Physiology at the Faculty of Medicine, Akita University, the former of whom responded "I am very familiar with it and/or an expert." and the latter of whom responded "I am familiar with it and/or a candidate for an expert." Meanwhile, the remainder who responded "I am not at all familiar with it." were filtered out. More specifically, a human subject was randomly offered one of 120 pairs of 15 averaged fluorescent cell images and selected the 11-grade similarity for each pair: from "10: 100% (perfectly similar/matched)" to "0: 0%." Note that, as a result, two accepted human subjects precisely evaluated "10: 100% (perfectly similar/matched)" for any pair of each of the 15 averaged fluorescent cell images and itself.

5.2. Experiment I

Experiment I shows the experimental results to validate the proposed method for automatic detection of bio-molecule clusters in a fluorescent cell image (as described in Section 3) using Dataset I.

Figure 12 shows the mAP @ IoU and F1-score @ IoU of the proposed method of Step 4(c) and Step 4(d) flagged as "1st" or "3rd" by manually optimizing the size of kernel, `kernel_size`, for OpenCV's Laplacian operator [69], `cv2.Laplacian()`, at Step 2(b), and Figure 13 shows each example of automatically detected bio-molecule clusters by the proposed methods of Step 4(c) and Step 4(d) flagged as "1st" or "3rd" of an input image. In addition, Table 2 compares the mAP @ IoU = 0.5 and F1-score @ IoU = 0.5 of the proposed methods of Step 4(c) and Step 4(d) flagged as "1st" or "3rd" for each input of

15 average fluorescent cell images. An analysis of these figures and table provides the following findings:

- The proposed method of Step 4(c) performs not low with respect to F1-score @ IoU, while it performs too low with respect to mAP @ IoU (i.e., precision @ IoU).
- The proposed method of Step 4(d) flagged as "1st" performs the best with respect to mAP @ IoU, while it performs too low with respect to F1-score @ IoU (i.e., recall @ IoU).
- The proposed method of Step 4(d) flagged as "3rd" performs the best with respect to F1-score @ IoU and also performs not low with respect to mAP @ IoU.
- The particles detected by the Mosaic Particle Tracker 2D/3D [13] can be only circular and uniform in size (e.g., `radius` = 3), while the proposed method could automatically detect bio-molecule clusters that are not only circular and not always uniform in size, as shown in Figures 3 and 13.
- F1-score @ IoU of the proposed method is lower than mAP @ IoU. More specifically, the recall @ IoU is worse than the precision @ IoU. It seems to be caused by over-filtering of Step 4(d) and the limitations of Dataset I; e.g., the ground truth is based on the particles detected by the Mosaic Particle Tracker 2D/3D [13], which can be only circular and uniform in size , while the proposed method could automatically detect bio-molecule clusters that are not only circular and not always uniform in size. The future work will make the dataset larger and more ground-true.

Therefore, this paper has concluded that the proposed method of Step 4(d) flagged as "3rd" is the best for automatic detection of bio-molecule clusters in a fluorescent cell image using Dataset I.

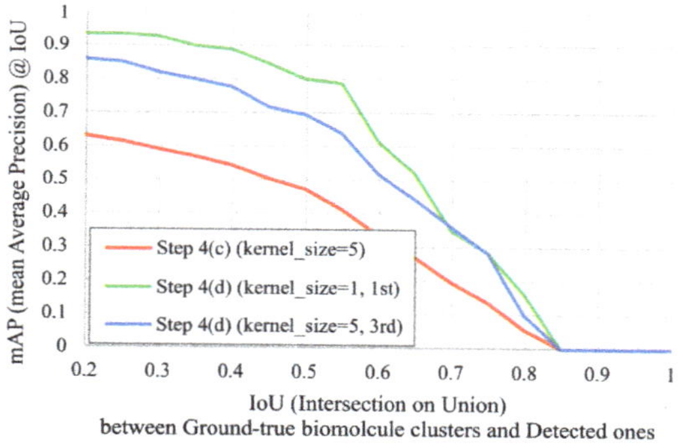

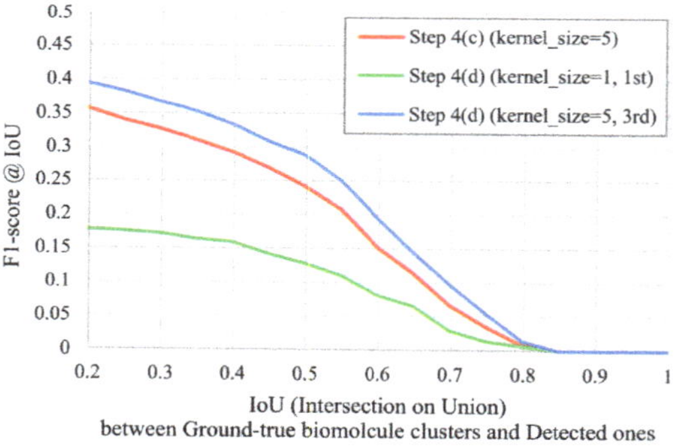

Figure 12. The mAP @ IoU and F1-score @ IoU of the proposed methods of Step 4(c) and Step 4(d) flagged as "1st" or "3rd" by manually optimizing the size of kernel, `kernel_size`, for OpenCV's Laplacian operator [69], `cv2.Laplacian()`, at Step 2(b).

(**a**)

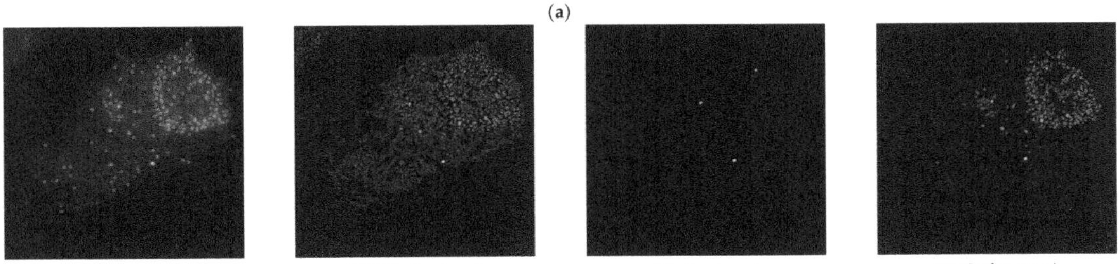

(177 clusters) Ground Truth | (1305 clusters) Step 4(c) | (3 clusters) Step 4(d) w "1st" | (300 clusters) Step 4(d) w "3rd"

(**b**)

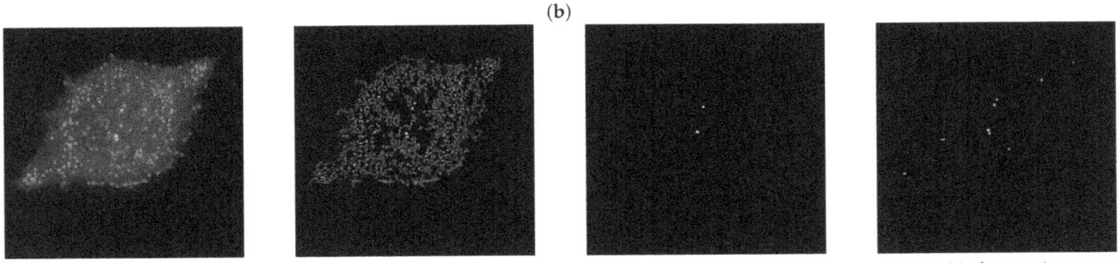

(210 clusters) Ground Truth | (902 clusters) Step 4(c) | (2 clusters) Step 4(d) w "1st" | (9 clusters) Step 4(d) w "3rd"

Figure 13. The ground truth, Step 4(c), and Step 4(d) flagged as "1st" or "3rd" of an input image (fluorescent cell #6 or #14) for automatic detection of bio-molecule clusters with the size of kernel, `kernel_size` = 1, for OpenCV's Laplacian operator [69], `cv2.Laplacian()`, at Step 2(b). (**a**) Cell #6; (**b**) Cell #14.

Table 2. The mAP @ IoU = 0.5 and F1-score @ IoU = 0.5 of the proposed methods of Steps 4(c) and 4(d) flagged as "1st" or "3rd" for each input of 15 average fluorescent cell images with the size of kernel, `kernel_size` = 5, for OpenCV's Laplacian operator [69], `cv2.Laplacian()`, at Step 2(b).

Cell #	Step 4(c) mAP	Step 4(c) F1-Score	Step 4(d) w "1st" mAP	Step 4(d) w "1st" F1-Score	Step 4(d) w "3rd" mAP	Step 4(d) w "3rd" F1-Score
Cell #1	0.288	0.186	1.000	0.019	1.000	0.019
Cell #2	0.604	0.284	0.741	0.114	0.698	0.460
Cell #3	0.309	0.140	0.554	0.130	0.554	0.127
Cell #4	0.385	0.166	0.609	0.152	0.491	0.229
Cell #5	0.364	0.261	0.833	0.019	0.506	0.119
Cell #6	0.478	0.135	1.000	0.044	0.584	0.411
Cell #7	0.440	0.177	0.806	0.031	0.747	0.041
Cell #8	0.573	0.256	0.735	0.300	0.641	0.458
Cell #9	0.593	0.216	0.686	0.055	0.695	0.423
Cell #10	0.424	0.192	0.975	0.099	0.921	0.122
Cell #11	0.536	0.283	1.000	0.015	0.639	0.404
Cell #12	0.393	0.233	0.729	0.069	0.729	0.069
Cell #13	0.491	0.337	0.833	0.028	0.618	0.283
Cell #14	0.603	0.300	1.000	0.037	0.899	0.123
Cell #15	0.609	0.415	0.565	0.088	0.698	0.458
Avg. (μ)	0.472	0.239	**0.804**	0.080	0.695	**0.250**
SD (σ)	0.107	0.075	0.157	0.072	0.144	0.165

Figure 14 shows the dependency of the mAP @ IoU and F1-score @ IoU of the proposed method of Step 4(d) flagged as "3rd" and $n = 5$ (set as the default for the number of sampled histograms for Step 4(d) in this paper) on the size of kernel, `kernel_size`, for OpenCV's

Laplacian operator [69], `cv2.Laplacian()`, at Step 2(b), respectively. An analysis of the figures provides the following findings:

- The dependency of mAP @ IoU on the size of kernel is more stable, while the dependency of F1-score @ IoU on the size of kernel is less stable. More specifically, the dependency of recall @ IoU on the size of kernel is less stable than the dependency of precision @ IoU on the size of kernel. It might be caused by the limitations of Dataset I. The future work will make the dataset larger and more ground-true.
- The curve of mAP over IoU is the best when the size of kernel is set to 1 and the 2nd best when the size of kernel is set to 5, and then, the larger the size of kernel is, the slightly worse the curve of mAP over IoU is.
- The curve of F1-score over IoU is the best when the size of kernel is set to 5 and the 2nd best when the size of kernel is set to 3, and then, the larger the size of kernel is, the worse the curve of F1-score over IoU is.
- Overall, the curves of both mAP over IoU and F1-score over IoU come in a slamming 1st place when the size of kernel is set to 5.

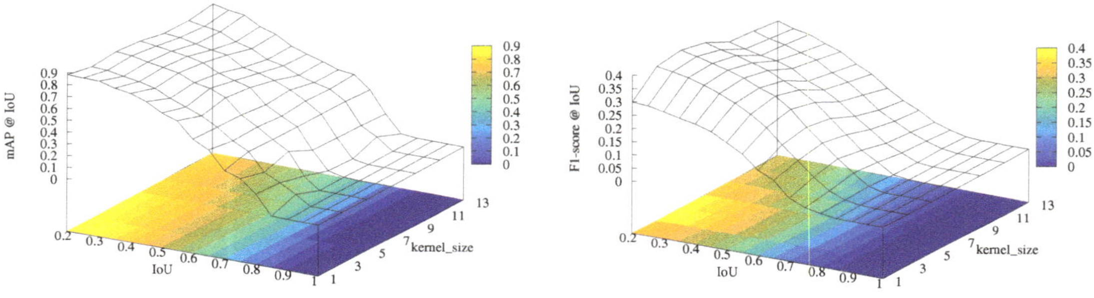

Figure 14. The mAP and F1-score @ IoU of Step 4(d) flagged as "3rd" and $n = 5$ depend on the size of kernel, `kernel_size`, for OpenCV's Laplacian operator [69], `cv2.Laplacian()`, at Step 2(b).

Figure 15 shows the dependency of the mAP @ IoU and F1-score @ IoU of the proposed methods of Step 4(d) flagged as "3rd" and `kernel_size` = 5 (which is manually optimized as the overall finding of the dependency analysis of mAP @ IoU and F1-score @ IoU on the size of kernel) on the number of sampled histograms, n, for Step 4(d), respectively. An analysis of the figures provides the following findings:

- The dependency of mAP @ IoU on the number of sampled histograms is more stable, while the dependency of F1-score @ IoU on the number of sampled histograms is less stable. More specifically, the dependency of recall @ IoU on the number of sampled histograms is less stable than the dependency of precision @ IoU on the number of sampled histograms. This might be caused by the limitations of Dataset I. The future work will make the dataset larger and more ground-true.
- The larger the number of sampled histograms is, the slightly worse the curve of mAP over IoU is. Note that it seems to converge.
- The larger the number of sampled histograms is, the better the curve of F1-score over IoU is. Note that it seems to converge.
- Overall, the curves of both mAP over IoU and F1-score over IoU come in 1st place when the number of sampled histograms is set to 5 as the default in this paper. Note that, the larger the number n of sampled histograms is, the greater the computation time for sampling n kinds of histograms is.

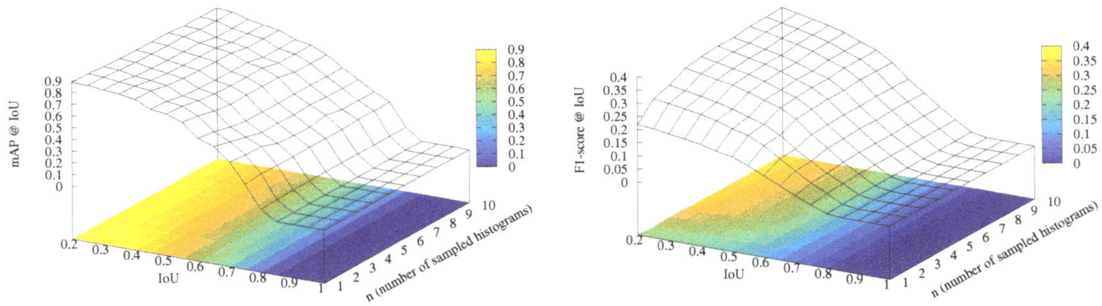

Figure 15. The mAP and F1-score @ IoU of Step 4(d) flagged as "3rd" and `kernel_size` = 5 depend on the number of sampled histograms, n, for Step 4(d).

Finally, Figure 16 shows the mAP @ IoU and F1-score @ IoU of the proposed methods of from Step 3(d) and Step 4(c) by manually optimizing the size of kernel, `kernel_size`, for OpenCV's Laplacian operator [69], `cv2.Laplacian()`, at Step 2(b). The proposed method of Step 4(c) is superior to its following methods from Step 3(d) to Step 4(b); i.e., Step 4(a) to Step 4(c) as well as Step 4(d) have good effects on automatic detection of bio-molecule clusters in a fluorescent cell image using Dataset I.

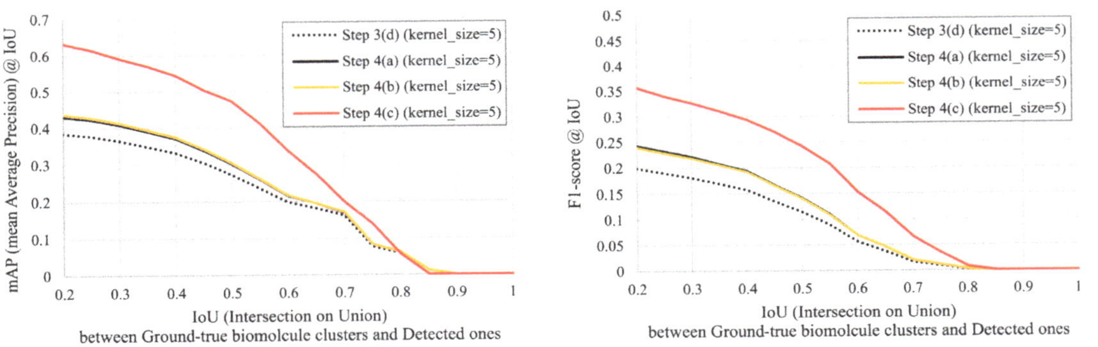

Figure 16. The mAP @ IoU and F1-score @ IoU of the proposed methods from Step 3(d) to Step 4(c) by manually optimizing the size of kernel, `kernel_size`, for OpenCV's Laplacian operator [69], `cv2.Laplacian()`, at Step 2(b).

Note that the pre-trained models of YOLOv8 [72] on the COCO dataset and ImageNet dataset, which is a state-of-the-art object detection for general purposes, cannot detect any bio-molecule clusters in an input TIRF image of a fluorescent cell. To achieve good performance while avoiding experts' (i.e., supervisors') biases, the existing AI technologies based on supervised ML (Machine Learning) or DL (Deep Learning) specific to the practical purpose of this paper need a larger dataset of TIRF images (maybe at least 1000 images) of fluorescent cells and their ground truth of bio-molecule clusters manually annotated by as many experts as possible. This is too expensive and takes too much time. Meanwhile, the proposed method does not need any large dataset for pre-training but only needs heuristics and statistics.

5.3. Experiment II

Experiment II shows the experimental results to validate the proposed SimMolCC, a similarity of automatically detected bio-molecule clusters between fluorescent cells (as described in Section 4) using Dataset II.

It has been finally found that ratio_area_BB at Step 3(d) (kernel_size = 5) provides the best Pearson Correlation Coefficient with two human subjects' 11-grade similarity (i.e., ground truth in the Dataset II). Figure 17 compares the Pearson Correlation Coefficient between two human subjects' 11-grade similarity and the proposed SimMolCC by cosine-similarity between two vectors of input images of a fluorescent cell based on the following 4 kinds features (i.e., histograms) of its automatically detected bio-molecule clusters at Step 3(d), and shows their dependency on the size of kernel, kernel_size, for OpenCV's Laplacian operator [69], cv2.Laplacian(), at Step 2(b):

- area: The histogram of area of each automatically detected bio-molecule cluster, where range = $(0, 200)$ and bins = 40.
- intensity: The histogram of mean fluorescence intensity of each automatically detected bio-molecule cluster, where range = $(0, 4096)$ and bins = 64.
- ratio_area_BB: the histogram of ratio of area to Bounding Box of each automatically detected bio-molecule cluster, where range = $(0.0, 1.0)$ and bins = 50.
- ratio_width_height: the histogram of ratio of width to height or ratio of height to width, whichever is smaller, of Bounding Box of each automatically detected bio-molecule cluster, where range = $(0.0, 1.0)$ and bins = 50.

Figure 17 also compares the Pearson Correlation Coefficient between two subjects' 11-grade similarity and the proposed SimMolCC by cosine-similarity between two vectors of input images of a fluorescent cell based on ratio_area_BB at Step 3(d), Step 4(c), and Step 4(d) (3rd), and shows their dependency on the size of kernel, kernel_size, for OpenCV's Laplacian operator [69], cv2.Laplacian(), at Step 2(b).

An analysis of the figures has found the following:

- ratio_area_BB at Step 3(d) (kernel_size = 5) provides the best Pearson Correlation Coefficient with two human subjects' similarity (i.e., ground truth in the Dataset II) and could help experts to conduct deeper analyses of bio-molecule clusters in a TIRF image of a fluorescent cell as their global features (not local features).
- ratio_area_BB (and ratio_width_height) of our proposed SimMolCC can represent the "shape" of each automatically detected bio-molecule cluster with not a uniform size, while Mosaic Particle Tracker 2D/3D [13], which is one of the most conventional methods for experts, can detect only circular one with a uniform size (e.g., radius = 3 of the target particles, meaning that the area is uniformly 29 [pixels]).
- Meanwhile, intensity provides too low Pearson Correlation Coefficient with two human subjects' similarity, independent of the size of kernel, kernel_size, for OpenCV's Laplacian operator [69], cv2.Laplacian(), at Step 2(b).

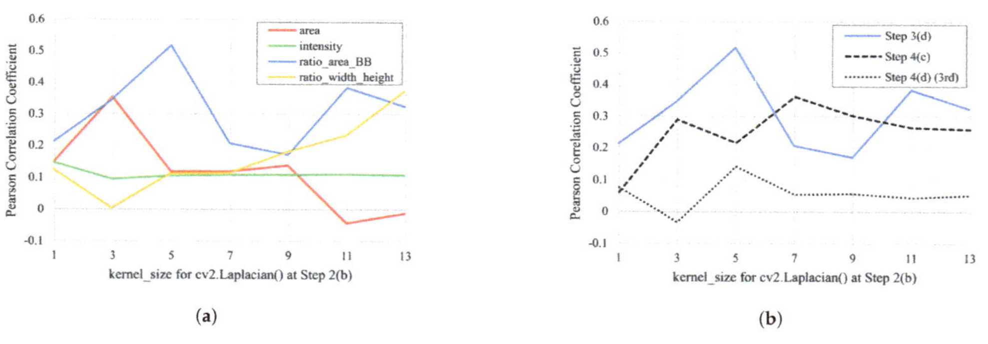

Figure 17. The Pearson Correlation Coefficient between two human subjects' 11-grade similarity and the proposed similarity, SimMolCC, depends on the size of kernel, kernel_size, for OpenCV's Laplacian operator [69], cv2.Laplacian(), at Step 2(b). (**a**) A comparison between histograms when Step 3(d) is constantly adopted. (**b**) A comparison between steps when ratio_area_BB is constantly adopted.

Figure 18 shows the scatter plot of two human subjects' 11-grade similarity and our proposed SimMolCC by cosine-similarity between two vectors of input images of a fluorescent cell based on `ratio_area_BB` at Step 3(d) (`kernel_size` = 5) for each of 105 (= ${}_{15}C_2$) pairs between the above-mentioned 15 averaged fluorescent cell images, and also shows the scatter plot of two human subjects' similarity and the converted SimMolCC′ from our proposed SimMolCC by the following formula:

$$\text{SimMolCC}'(img_1, img_2) := \frac{1}{10} \cdot (563.18 \cdot \text{SimMolCC}(img_1, img_2) - 554.67) \quad (4)$$

where $y = 563.18 \cdot x - 554.67$ has been obtained by simple linear regression from SimMolCC (as x) for two human subjects' 11-grade similarity (as y).

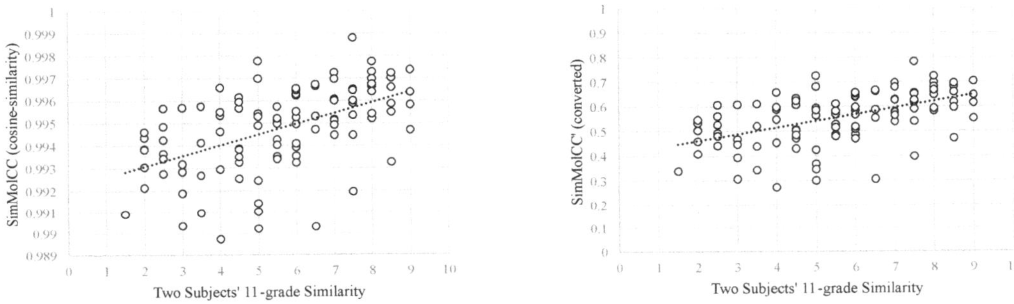

Figure 18. The scatter plots of two human subjects' 11-grade similarity and the proposed SimMolCC, or the converted SimMolCC′ from the proposed SimMolCC by simple linear regression.

Finally, Figure 19 shows an example result of similarity-based retrieving (ranking) by inputting a TIRF image (fluorescent cell #14) as a query and calculating its SimMolCC′ with the other 14 TIRF images. The ranking based on the converted SimMolCC′ from the proposed SimMolCC by simple linear regression has achieved similar results as the ranking based on two human subjects' 11-grade similarity.

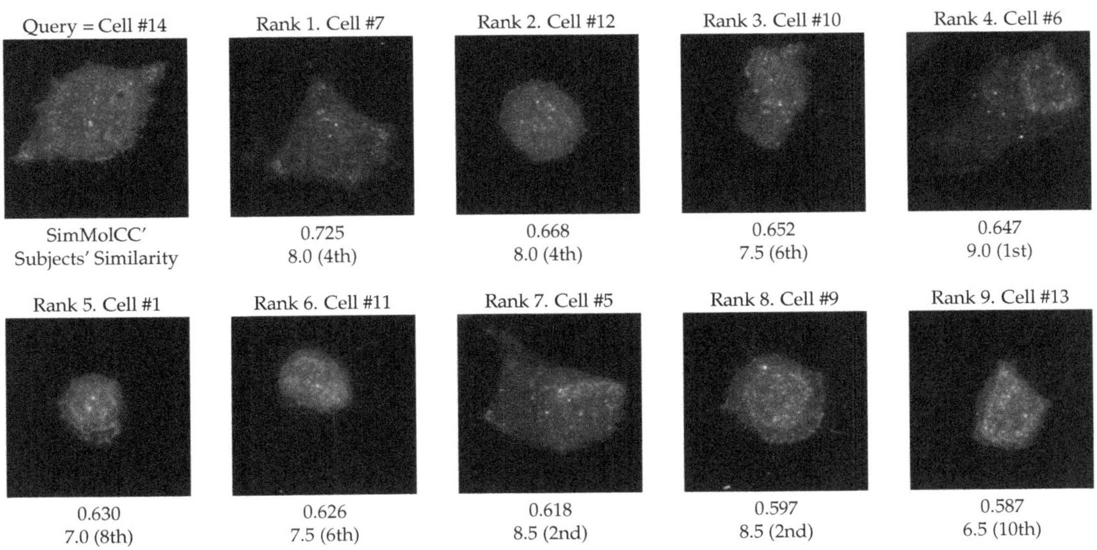

Figure 19. *Cont.*

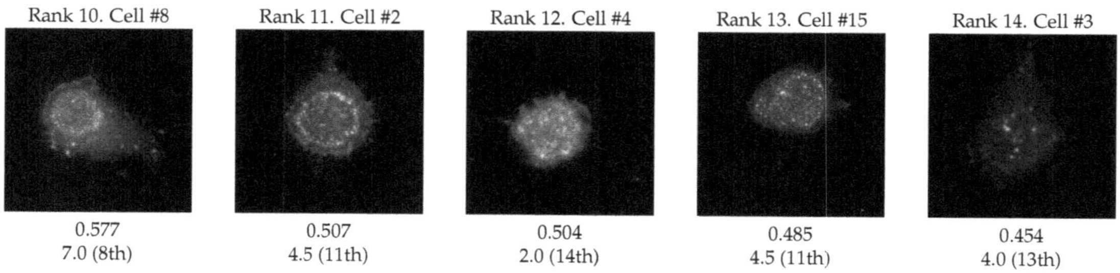

Figure 19. An example result of similarity-based retrieval (ranking) by inputting a TIRF image (fluorescent cell #14) as a query and calculating its SimMolCC′ with the other 14 TIRF images.

6. Conclusions

In the field of studies on the "Neural Synapses" in the nervous system, its experts manually (or pseudo-automatically) detect bio-molecule clusters (e.g., of proteins) in many TIRF (Total Internal Reflection Fluorescence) images of a fluorescent cell and analyze their static/dynamic behaviors. This paper has proposed a novel method for the automatic detection of the bio-molecule clusters in a TIRF image of a fluorescent cell and conducted several experiments on its performance, e.g., mAP @ IoU (mean Average Precision @ Intersection over Union) and F1-score @ IoU, as an objective/quantitative means of evaluation. As a result, the best of the proposed methods has achieved 0.695 as its mAP @ IoU = 0.5 and 0.250 as its F1-score @ IoU = 0.5 and would have to be improved, especially with respect to its recall @ IoU. But, the proposed method could automatically detect bio-molecule clusters that are not only circular and not always uniform in size, and can output various histograms and heatmaps for novel deeper analyses of the automatically detected bio-molecule clusters, while the particles detected by the Mosaic Particle Tracker 2D/3D [13], which is one of the most conventional methods for experts, can be only circular and uniform in size.

In addition, this paper has defined and validates a novel similarity of automatically detected bio-molecule clusters between fluorescent cells, i.e., SimMolCC. As a result, the best of the proposed methods has achieved 0.518 (p-value < 0.001, statistically significant [73,74]) as its Pearson Correlation Coefficient with two human subjects' 11-grade similarity, which would have to be improved in the future. But, the findings include that the histogram of the ratio of area to Bounding Box, `ratio_area_BB`, of each automatically detected bio-molecule cluster is superior to the histogram of its `intensity` as its global features help experts to conduct deeper analyses of the bio-molecule clusters in a TIRF image of a fluorescent cell; i.e., the "shape" of each automatically detected bio-molecule cluster with a non-uniform size plays an important role in novel deeper analyses by experts.

In the near future, the implemented tools with the proposed method will be developed for experts and applied in various studies on "Neural Synapses" for more advances in both Brain Science and Artificial Neural Networks. In addition, the future work includes validating the other definitions of SimMolCC based on the six kinds of heatmaps and also various hybrids with some of the four kinds of histograms, e.g., a hybrid of `ratio_area_BB` at Step 3(d) with `kernel_size` $= 5$ and `ratio_width_height` at Step 3(d) with `kernel_size` $= 13$, and/or some of the six kinds of heatmaps, such as the global feature vector $\vec{v}_i$ of each input TIRF image of a fluorescent cell, with or without converting their frequencies of a global feature vector $\vec{v}_i$ by the $log_2()$ function, like TF–IDF (Term Frequency–Inverse Document Frequency).

Author Contributions: Conceptualization, S.H. and M.T.; methodology, S.H.; software, S.H.; validation, S.H.; formal analysis, S.H.; investigation, S.H., M.T. and T.M.; resources, T.M., A.S. and D.K.; data curation, S.H. and A.S.; writing—original draft, S.H.; writing—review and editing, S.H., T.M., A.S., D.K. and M.T.; visualization, S.H. and M.T.; supervision, T.M.; project administration, S.H.; funding acquisition, S.H. and T.M. All authors have read and agreed to the published version of the manuscript.

Funding: This research was funded by JSPS's (the Japan Society for the Promotion of Science) KAKENHI grants (24K06287 to S.H.; 21H02584 to T.M.).

Institutional Review Board Statement: All procedures and animal care were conducted in accordance with the guidelines of the Physiological Society of Japan and were approved by Akita University committee for Regulation on the Conduct of Animal Experiments and Related Activities (approval number: a-1-505; date of approval: 2 May 2023).

Informed Consent Statement: Informed consent was obtained from all subjects involved in the study.

Data Availability Statement: The datasets presented in this article might not be available on request from the corresponding author due to ethical reasons of animals and privacy reasons of subjects.

Acknowledgments: This work was partially supported by Regional ICT Research Center of Human, Industry and Future at The University of Shiga Prefecture, and by Cabinet Office, Government of Japan.

Conflicts of Interest: The authors declare no conflicts of interest.

References

1. Society 5.0. Available online: https://www8.cao.go.jp/cstp/english/society5_0/index.html (accessed on 8 August 2024).
2. Hattori, S.; Aiba, K.; Takahara, M. R2-B2: A Metric of Synthesized Image's Photorealism by Regression Analysis based on Recognized Objects' Bounding Box. In Proceedings of the Joint 12th International Conference on Soft Computing and Intelligent Systems and 23rd International Symposium on advanced Intelligent Systems (SCIS&ISIS'22), Online/Ise-Shima, Mie, Japan, 29 November–2 December 2022; F-1-F-1. [CrossRef]
3. Hattori, S.; Takahara, M. A Study on Human-Computer Interaction with Text-to/from-Image Game AIs for Diversity Education. In Proceedings of the 25th International Conference on Human-Computer Interaction (HCI International 2023), Online/Copenhagen, Denmark, 23–28 July 2023; LNCS. Volume 14015, pp. 471–486. [CrossRef]
4. Takahara, M.; Hattori, S. A Study on HCI of a Collaborated Nurture Game for Sleep Education with Child and Parent. In Proceedings of the 25th International Conference on Human-Computer Interaction (HCI International 2023), Online/Copenhagen, Denmark, 23–28 July 2023; LNCS. Volume 14015, pp. 169–181. [CrossRef]
5. Takahara, M.; Nishimura, S.; Hattori, S. A Study on a Mechanism to Prevent Sleeping Smartphones using ASMR. In Proceedings of the 26th International Conference on Human-Computer Interaction (HCI International 2024), Online/Washington DC, USA, 29 June 29–4 July 2024; LNCS. Volume 14689, pp. 279–288. [CrossRef]
6. Hattori, S.; Miyamoto, S.; Sunayama, W.; Takahara, M. A Study on Input Methods of User Preference for Personalized Fashion Coordinate Recommendations. In Proceedings of the 26th International Conference on Human-Computer Interaction (HCI International 2024), Online/Washington DC, USA, 29 June 29–4 July 2024; LNCS. Volume 14691, pp. 178–196. [CrossRef]
7. SAMOE—Simple Simulation for Semi-Order Made Apron -Normal Pattern-. Available online: https://samoe.net/f/simulation-normal (accessed on 8 August 2024).
8. Arasawa, K.; Hattori, S. Automatic Baseball Video Tagging based on Voice Pattern Prioritization and Recursive Model Localization. *J. Adv. Comput. Intell. Intell. Inform.* **2017**, *21*, 1262–1279. [CrossRef]
9. Watanabe, R.; Arasawa, K.; Hattori, S. Rule-Based Role Analysis of Game Characters Using Tags about Characteristics for Strategy Estimation by Game AI. In Proceedings of the Intelligent Systems Workshop 2018 (ISWS '18) in Conjunction with SCIS&ISIS'18, Toyama, Japan, 5–8 December 2018; Fr6-1-5; pp. 814–819.
10. Hattori, S.; Kurono, M.; Yoshida, Y.; Takahara, M.; Kudo, Y. Time Control of Thinking and Cursor Movement for Humanized Othello AIs. *Inf. Process. Soc. Jpn. Trans. Database* **2023**, *16*, 16–33
11. Yang, Z.R.; Yang, Z. Chapter 6.01—Artificial Neural Networks. In *Comprehensive Biomedical Physics*; Volume 6: Bioinformatics; Persson, B., Ed.; Elsevier: Amsterdam, Netherlands, 2014; pp. 1–17. [CrossRef]
12. Cheezum, M.K.; Walker, W.F.; Guilford, W.H. Quantitative comparison of algorithms for tracking single fluorescent particles. *Elsevier Biophisical J.* **2001**, *81*, 2378–2388. [CrossRef] [PubMed]
13. 3.1. Particle Tracker 2D/3D—MosaicSuite 1.0.23 documentation. Available online: https://sbalzarini-lab.org/MosaicSuiteDoc/particleTracker.html (accessed on 8 August 2024).
14. Miki, T.; Hashimotodani, Y.; Sakaba, T. Chapter 13—Synaptic vesicle dynamics at the calyx of Held and other central synapses. In *Exocytosis from Molecules to Cells*; Anantharam, A., Knight, J., Eds.; IOP Publishing: Bristol, England, 2022; pp. 13-1–13-18. [CrossRef]
15. Zenisek, D.; Steyer, J.; Almers, W. Transport, capture and exocytosis of single synaptic vesicles at active zones. *Nature* **2000**, *406*, 849–854. [CrossRef]
16. Miki, T.; Midorikawa, M.; Sakaba, T. Direct imaging of rapid tethering of synaptic vesicles accompanying exocytosis at a fast central synapse. *Proc. Natl. Acad. Sci. USA* **2020**, *117*, 14493–14502. [CrossRef] [PubMed]
17. Ultralytics YOLOv8 Docs. Available online: https://docs.ultralytics.com/ (accessed on 7 April 2024).

18. Kaur, J.; Singh, W. A systematic review of object detection from images using deep learning. *Multimed. Tools Appl.* **2024**, *83*, 12253–12338. [CrossRef]
19. Chen, W.; Luo, J.; Zhang, F.; Tian, Z. A review of object detection: Datasets, performance evaluation, architecture, applications and current trends. *Multimed. Tools Appl.* **2024**, *83*, 65603–65661. [CrossRef]
20. Vijayakumar, A.; Vairavasundaram, S. YOLO-based Object Detection Models: A Review and its Applications. *Multimed. Tools Appl.* **2024**. [CrossRef]
21. Sinha, P.K.; R, M. Conglomeration of deep neural network and quantum learning for object detection: Status quo review. *Knowl.-Based Syst.* **2024**, *288*, 111480. [CrossRef]
22. Wei, W.; Cheng, Y.; He, J.; Zhu, X. A review of small object detection based on deep learning. *Neural Comput. Appl.* **2024**, *36*, 6283–6303. [CrossRef]
23. Flores-Calero, M.; Astudillo, C.A.; Guevara, D.; Maza, J.; Lita, B.S.; Defaz, B.; Ante, J.S.; Zabala-Blanco, D.; Armingol Moreno, J.M. Traffic Sign Detection and Recognition Using YOLO Object Detection Algorithm: A Systematic Review. *Mathematics* **2024**, *12*, 297. [CrossRef]
24. Zhao, R.; Tang, S.; Supeni, E.E.B.; Rahim, S.B.A.; Fan, L. A Review of Object Detection in Traffic Scenes Based on Deep Learning. *Sciendo Appl. Math. Nonlinear Sci.* **2024**, *9*. [CrossRef]
25. Tahir, N.U.A.; Zhang, Z.; Asim, M.; Chen, J.; ELAffendi, M. Object Detection in Autonomous Vehicles under Adverse Weather: A Review of Traditional and Deep Learning Approaches. *Algorithms* **2024**, *17*, 103. [CrossRef]
26. Song, S.; Liu, L.; Jia, F.; Luo, Y.; Zhang, G.; Yang, L.; Wang, L.; Jia, C. Robustness-Aware 3D Object Detection in Autonomous Driving: A Review and Outlook. *IEEE Trans. Intell. Transp. Syst.* **2024**. [CrossRef]
27. Yang, D.; Solihin, M.I.; Zhao, Y.; Yao, B.; Chen, C.; Cai, B.; Machmudah, A. A review of intelligent ship marine object detection based on RGB camera. *IET Image Process.* **2024**, *18*, 281–297. [CrossRef]
28. Zhao, C.; Liu, R.W.; Qu, J.; Gao, R. Deep learning-based object detection in maritime unmanned aerial vehicle imagery: Review and experimental comparisons. *Elsevier Eng. Appl. Artif. Intell.* **2024**, *128*, 107513. [CrossRef]
29. Goyal, V.; Singh, R.; Dhawley, M.; Kumar, A.; Sharma, S. Aerial Object Detection Using Deep Learning: A Review. *Comput. Intell. Sel. Proc. InCITe* **2022**, *2023*, 81–92. [CrossRef]
30. Gui, S.; Song, S.; Qin, R.; Tang, Y. Remote Sensing Object Detection in the Deep Learning Era—A Review. *Remote. Sens.* **2024**, *16*, 327. [CrossRef]
31. Ariza-Sentís, M.; Vélez, S.; Martínez-Peñ, R.; Baja, H.; Valente, J. Object detection and tracking in Precision Farming: A systematic review. *Comput. Electron. Agric.* **2024**, *219*, 108757. [CrossRef]
32. Badgujar, C.M.; Poulose, A.; Gan, H. Agricultural Object Detection with You Look Only Once (YOLO) Algorithm: A Bibliometric and Systematic Literature Review. *arXiv* **2024**, arXiv:2401.10379. [CrossRef]
33. Sharma, P.; Saurav, S.; Singh, S. Object detection in power line infrastructure: A review of the challenges and solutions. *Eng. Appl. Artif. Intell.* **2024**, *130*, 107781. [CrossRef]
34. O'Connor, M.F.; Hughes, A.; Zheng, C.; Davies, A.; Kelleher, D.; Ahmad, K. Annotation and Retrieval of Cell Images. In Proceedings of the Intelligent Data Engineering and Automated Learning (IDEAL'10), Paisley, UK, 1–3 September 2010; LNCS Volume 6283, pp. 218–225. [CrossRef]
35. Koprowski, R.; Wrobel, Z. Automatic segmentation of biological cell structures based on conditional opening and closing. *Mach. Graph. Vis. Int. J.* **2005**, *14*, 285–307.
36. Han, J.W.; Breckon, T.P.; Randell, D.A.; Landini, G. The application of support vector machine classification to detect cell nuclei for automated microscopy. *Mach. Vis. Appl.* **2012**, *23*, 15–24. [CrossRef]
37. Barbu, T. SVM-based Human Cell Detection Technique using Histograms of Oriented Gradients. *Math. Methods Inf. Sci. Econ.* **2012**, *4*, 156–160.
38. Mualla, F.; Scholl, S.; Sommerfeldt, B.; Maier, A.; Hornegger, J. Automatic Cell Detection in Bright-Field Microscope Images Using SIFT, Random Forests, and Hierarchical Clustering. *IEEE Trans. Med. Imaging* **2013**, *32*, 2274–2286. [CrossRef] [PubMed]
39. Öztürk, Ş.; Bayram, A. Comparison of HOG, MSER, SIFT, FAST, LBP and CANNY features for cell detection in histopathological images. *Helix* **2018**, *8*, 3321–3325. [CrossRef]
40. Akram, S.U.; Kannala, J.; Eklund, L.; Heikkilä, J. Cell Segmentation Proposal Network for Microscopy Image Analysis. In Proceedings of the International Workshop on Deep Learning and Data Labeling for Medical Applications (DLMIA'16 and LABELS'16), Athens, Greece, 21 October 2016; LNCS. Volume 10008, pp. 21–29. [CrossRef]
41. Al-Kofahi, Y.; Zaltsman, A.; Graves, R.; Marshall, W.; Rusu, M. A deep learning-based algorithm for 2-D cell segmentation in microscopy images. *BMC Bioinform.* **2018**, *19*, 365. [CrossRef] [PubMed]
42. Pachitariu, M.; Stringer, C. Cellpose 2.0: How to train your own model. *Nat. Methods* **2022**, *19*, 1634–1641. [CrossRef]
43. Zhang, Z.; Liu, Q.; Wang, Y. Road Extraction by Deep Residual U-Net. *IEEE Geosci. Remote. Sens. Lett.* **2018**, *15*, 749–753. [CrossRef]
44. Ghaznavi, A.; Rychtáriková, R.; Saberioon, M.; Štys, D. Cell segmentation from telecentric bright-field transmitted light microscopy images using a Residual Attention U-Net: A case study on HeLa line. *Comput. Biol. Med.* **2022**, *147*, 105805. [CrossRef]
45. Ronneberger, O.; Fischer, P.; Brox, T. U-Net: Convolutional Networks for Biomedical Image Segmentation. In Proceedings of the 18th International Conference on Medical Image Computing and Computer-Assisted Intervention (MICCAI'15), Munich, Germany, 5–9 October 2015; LNCS. Volume 9351, pp. 234–241. [CrossRef]

46. He, K.; Zhang, X.; Ren, S.; Sun, J. Deep Residual Learning for Image Recognition. In Proceedings of the 2016 IEEE Conference on Computer Vision and Pattern Recognition (CVPR'16), Las Vegas, NV, USA, 26 June–1 July 2016; pp. 770–778. [CrossRef]
47. Alom, M.Z.; Hasan, M.; Yakopcic, C.; Taha, T.M.; Asari, V.K. Recurrent Residual Convolutional Neural Network based on U-Net (R2U-Net) for Medical Image Segmentation. *arXiv* **2018**, arXiv:1802.06955. [CrossRef]
48. Xing, F.; Yang, L. Robust Nucleus/Cell Detection and Segmentation in Digital Pathology and Microscopy Images: A Comprehensive Review. *IEEE Rev. Biomed. Eng.* **2016**, *9*, 234–263. [CrossRef]
49. Mustafa, W.A.; KADER, M.M.M.A. A comparative study of automated segmentation methods for cell nucleus detection. *Malays. Appl. Biol.* **2018**, *47*, 125–129.
50. Ma, B.; Zhang, J.; Cao, F.; He, Y. MACD R-CNN: An Abnormal Cell Nucleus Detection Method. *IEEE Access* **2020**, *8*, 166658–166669. [CrossRef]
51. Shimomoto, Y.; Inoue, K.; Yamamoto, I.; Ohba, S.; Ogata, K.; Yamamoto, H. Cell Nucleus Detection in Oral Cytology Using Artificial Intelligence. *Sens. Mater.* **2023**, *35*, 399–409. [CrossRef]
52. Hashimoto, T.; Yamashita, K.; Yamazaki, K.; Hirayama, K.; Yabuzaki, J.; Kobayashi, H. Study of Analysis and Quantitative Estimation of Melanin in Face Epidermal Corneocyte. *Trans. Jpn. Soc. Mech. Eng. (JSME) C* **2012**, *78*, 508–522. [CrossRef]
53. Corneo Cytemetry SG (Second Generation). Available online: https://corneocytemetry.com/ (accessed on 8 August 2024).
54. Hasegawa, S.; Enomoto, K.; Mizutani, T.; Okano, Y.; Tanaka, T.; Sakai, O. Skin Diagnostic Method Using Fontana-Masson Stained Images of Stratum Corneum Cells. *IEICE Trans. Inf. Syst.* **2024**, *107*, 1070–1089. [CrossRef]
55. Hattori, S.; Tanaka, K. Search the Web for Peculiar Images by Converting Web-extracted Peculiar Color-Names into Color-Features. *IPSJ (Inf. Process. Soc. Jpn. Trans. Databases* **2010**, *3*, 49–63. http://id.nii.ac.jp/1001/00069213/.
56. Dowerah, R.; Patel, S. Comparative analysis of color histogram and LBP in CBIR systems. *Multimed. Tools Appl.* **2024**, *83*, 12467–12486. [CrossRef]
57. Lowe, D.G. Object recognition from local scale-invariant features. In Proceedings of the Seventh IEEE International Conference on Computer Vision (ICCV'99), Kerkyra, Greece, 20–27 September 1999; Volume 2, pp. 1150–1157. [CrossRef]
58. Bay, H.; Tuytelaars, T.; Van Gool, L. SURF: Speeded Up Robust Features. In Proceedings of the 9th European Conference on Computer Vision (ECCV'06), Graz, Austria, 7–13 May 2006; Part I. pp. 404–417. [CrossRef]
59. Cao, B.; Araujo, A.; Sim, J. Unifying Deep Local and Global Features for Image Search. In Proceedings of the 23th European Conference on Computer Vision (ECCV'20), Online, 23–28 August 2020; pp. 726–743. [CrossRef]
60. Li, L.; Che, D.; Wang, X.; Zhang, P.; Rahman, S.U.; Zhao, J.; Yu, J.; Tao, S.; Lu, H.; Liao, M. CellSim: A novel software to calculate cell similarity and identify their co-regulation networks. *BMC Bioinform.* **2019**, *20*, 111. [CrossRef]
61. Lin, D. An information-theoretic definition of similarity. In Proceedings of the Fifteenth International Conference on Machine Learning (ICML'98), Madison, WI, USA, 24–27 July 1998; pp. 296–304.
62. Lizio, M.; Harshbarger, J.; Abugessaisa, I.; Noguchi, S.; Kondo, A.; Severin, J.; Mungall, C.; Arenillas, D.; Mathelier, A.; Medvedeva, Y.A.; et al. Update of the FANTOM web resource: High resolution transcriptome of diverse cell types in mammals. *Nucleic Acids Res.* **2016**, *45*, D737–D743. [CrossRef] [PubMed]
63. Sato, S.; Fujieda, N.; Moriya, A.; Kise, K. SimCell: A Processor Simulator for Multi-Core Architecture Research. *Inf. Media Technol.* **2009**, *4*, 270–281. [CrossRef]
64. Fan, C.; Davidson, P.A.; Habgood, R.; Zeng, H.; Decker, C.M.; Salazar, M.G.; Lueangwattanapong, K.; Townley, H.E.; Yang, A.; Thompson, I.P.; et al. Chromosome-free bacterial cells are safe and programmable platforms for synthetic biology. *Proc. Natl. Acad. Sci. USA* **2020**, *117*, 6752–6761. [CrossRef] [PubMed]
65. Ostu, N. A Threshold Selection Method from Gray-Level Histograms. *IEEE Trans. Syst. Man Cybern.* **1979**, *9*, 62–66. [CrossRef]
66. Fang, M.; Yue, G.-X.; Yu, Q.-C. The Study on An Application of Otsu Method in Canny Operator. In Proceedings of the 2009 International Symposium on Information Processing (ISIP 09), Huangshan, China, 21–23 August 2009; pp. 109–112.
67. Sturges, H.A. The Choice of a Class Interval. *J. Am. Stat. Assoc.* **1926**, *21*, 65–66. [CrossRef]
68. Find_Peaks—SciPy v1.14.0 Manual. Available online: https://docs.scipy.org/doc/scipy/reference/generated/scipy.signal.find_peaks.html (accessed on 8 August 2024).
69. OpenCV: Laplace Operator. Available online: https://docs.opencv.org/4.x/d5/db5/tutorial_laplace_operator.html (accessed on 8 August 2024).
70. OpenCV: Canny Edge Detection. Available online: https://docs.opencv.org/4.x/da/d5c/tutorial_canny_detector.html (accessed on 8 August 2024).
71. Z-Functions. Available online: https://imagej.net/imaging/z-functions (accessed on 8 August 2024).
72. Detect-Ultralytics YOLO Docs. Available online: https://docs.ultralytics.com/tasks/detect/#models (accessed on 30 August 2024).
73. Pearson, E.S. The Test of Significance for the Correlation Coefficient. *J. Am. Stat. Assoc.* **1931**, *26*, 128–134. [CrossRef]
74. Pearson, E.S. The Test of Significance for the Correlation Coefficient: Some Further Results. *J. Am. Stat. Assoc.* **1932**, *27*, 424–426. [CrossRef]

applied sciences

MDPI

Article

Enhancing Detection of Pedestrians in Low-Light Conditions by Accentuating Gaussian–Sobel Edge Features from Depth Maps

Minyoung Jung and Jeongho Cho *

Department of Electrical Engineering, Soonchunhyang University, Asan 31538, Republic of Korea; jmy000321@naver.com
* Correspondence: jcho@sch.ac.kr; Tel.: +82-41-530-4960

Abstract: Owing to the low detection accuracy of camera-based object detection models, various fusion techniques with Light Detection and Ranging (LiDAR) have been attempted. This has resulted in improved detection of objects that are difficult to detect due to partial occlusion by obstacles or unclear silhouettes. However, the detection performance remains limited in low-light environments where small pedestrians are located far from the sensor or pedestrians have difficult-to-estimate shapes. This study proposes an object detection model that employs a Gaussian–Sobel filter. This filter combines Gaussian blurring, which suppresses the effects of noise, and a Sobel mask, which accentuates object features, to effectively utilize depth maps generated by LiDAR for object detection. The model performs independent pedestrian detection using the real-time object detection model You Only Look Once v4, based on RGB images obtained using a camera and depth maps preprocessed by the Gaussian–Sobel filter, and estimates the optimal pedestrian location using non-maximum suppression. This enables accurate pedestrian detection while maintaining a high detection accuracy even in low-light or external-noise environments, where object features and contours are not well defined. The test evaluation results demonstrated that the proposed method achieved at least 1–7% higher average precision than the state-of-the-art models under various environments.

Keywords: pedestrian detection; Gaussian–Sobel; depth map; low light; point cloud

1. Introduction

Autonomous driving technology has recently gained widespread acceptance among consumers and is becoming increasingly integrated into our daily lives, while making significant contributions to improving the quality of human life. It is projected to become a ubiquitous technology that will be easily accessible everywhere in the near future. Furthermore, it can overcome the physical or mental human limitations, improve safety and reliability, reduce the accident rate, and cut labor costs, thereby reducing social costs and increasing commercial value [1,2]. For autonomous driving in the mobility domain, a camera-based object detection system is fundamental for identifying lanes, vehicles/pedestrians, traffic signals, and other obstacles. However, in some environments, object detection becomes impossible owing to the difficulty of acquiring high-quality images using a camera. Moreover, object detection models that solely utilize visual data from cameras have exhibited considerably poor performance in pedestrian detection tasks, which is attributable to the inherent challenges in representing the diverse appearances of pedestrians within a unified shape [3,4].

Hsu and Yang [5] developed a two-stage pipeline to mitigate the inherent limitations in camera-based object detection systems. First, they leveraged Super-Resolution Generative Adversarial Networks to upscale low-resolution images obtained from cameras. The resulting high-resolution images were subsequently fed into a faster region-based convolutional neural network (Faster R-CNN) for pedestrian detection, demonstrating considerable enhancement in detection accuracy. Zhang et al. [6] improved pedestrian detection performance in occluded environments by suppressing feature extraction from object-free background images. Mushtaq et al. [7] proposed a single-image super-resolution network model based on CNNs, combining conventional autoencoders with residual neural network approaches. By removing the noise present in images, they improved the image quality and enhanced detection accuracy. Xu et al. [8] also proposed a method to enhance image quality by directly embedding a physical lighting model into a deep neural network to improve object detection performance in low-light conditions. This method captures the difference between the local content of an object and the preferred region of its local neighborhood. However, these techniques exhibit shortcomings in object detection when parts of objects are obscured by other objects, or when noise induced by light reflection, scattering, and other low-light conditions results in the loss of some contour information.

Given the limitations of camera-based object detection systems due to various external environmental factors such as low-light conditions, supplementary sensors have been incorporated into these systems to enhance their performance. Gilroy et al. [9] enhanced pedestrian detection performance in low-light conditions by generating a depth map based on Light Detection and Ranging (LiDAR) and stereovision sensors, mitigating the effects of light noise. Lin et al. [10] explored the use of CNNs to improve small object detection performance based on both three-dimensional (3D) point cloud data (PCD) and two-dimensional (2D) images. Qi et al. [11] predicted the location of objects in 2D images based on CNNs and converted the corresponding regions into three dimensions, thus improving the detection performance. Although all the aforementioned studies demonstrated enhanced detection accuracy of invisible objects by employing multiple sensors, their performance in detecting small objects in low-light environments and at a large distance from the sensor remained suboptimal, primarily owing to the challenges associated with object shape estimation in such environments. Although LiDAR was additionally used, object detection in low-light scenarios continued to be challenging because of the low-resolution depth maps produced by the LiDAR, which hindered accurate object detection. Hence, to improve the performance, supplementary image processing is essential to highlight the features of an object. A sharpening filter presents a viable option for this enhancement. Additionally, the PCD generated by LiDAR is susceptible to particle-based external noise, which is another factor to be considered when choosing a sharpening filter.

Over the years, researchers have explored various methods of image enhancement using sharpening filters. Maragos and Pessoa [12] achieved image enhancement by expanding pixels at the presumed locations of objects in all directions through morphological operations, thereby highlighting object contours. The image quality was enhanced by centering the structural elements on the foreground pixels. If a foreground pixel was present in the area where the elements overlapped, the central pixel of that area was designated as the foreground pixel, thereby enlarging the size and outline of the objects. Deng [13] enhanced the sharpness of objects by applying an unsharp mask filter that increases object density. This filter was created using a mask obtained from the difference between a blurred image and the original image, thereby increasing the contrast between pixels. Ali and Clausi [14] utilized a Canny edge filter to identify pixels with the greatest rate of change as edges and employed hysteresis edge tracking to reduce noise effects and eliminate isolated edges, thereby enhancing the sharpness of object contours. The application of reconstructed images using these filters to object detection models is expected to lead to considerable performance improvements. However, object detection performance could be degraded instead if some object information is missing owing to image filtering or if nearby objects appear to be overlapped. Furthermore, the inadvertent amplification of external noise may also adversely affect object detection performance.

To mitigate the aforementioned adverse effect, we propose a novel object detection model that employs a Gaussian–Sobel filter as a preprocessor. This filter effectively combines Gaussian blurring to suppress noise effects and the Sobel mask to accentuate object features, enabling effective object detection from LiDAR-derived depth maps. The proposed model employs RGB images and depth maps, preprocessed with the Gaussian–Sobel filter, to obtain respective detection results using the real-time object detection model, You Only Look Once (YOLO). The YOLO object detection model has been widely used to date. Moreover, new versions of the model are being continuously developed, which exhibit overall better performance in terms of speed and accuracy compared with other object detection models, including single-shot multibox detector (SSD) and Faster R-CNN. Furthermore, YOLO is a lightweight model that can be embedded into on-board systems with limited resources while maintaining a high detection performance. In particular, YOLOv4 is best suited for real-time object detection in real-world environments, as it enables real-time object detection through on-board systems and can detect objects, including small objects, in complex environments [15,16]. Optimal object locations are estimated by eliminating redundant bounding boxes through non-maximum suppression (NMS). This enables robust object detection, even under challenging conditions such as low-light environments or high-speckle-noise environments, by preserving a high detection accuracy despite the lack of distinct object features or contours. Our test evaluation results demonstrated that the proposed model outperformed existing state-of-the-art models by achieving a 1–7% improvement in average precision (AP), depending on the experimental conditions. The contributions of this study are as follows:

- We improved image resolution and enhanced object–background segmentation by preprocessing LiDAR-derived depth maps using a fusion of Gaussian blurring and the Sobel mask.
- We proposed a versatile object detection model that effectively combines RGB images from cameras and depth maps preprocessed by the Gaussian–Sobel filter. This convergence enables robust object detection in diverse lighting conditions, ranging from bright daylight to low-light environments, complementing the strengths and weaknesses of cameras and LiDAR.
- By applying the Gaussian–Sobel filter, we enhanced the robustness of LiDAR, leading to improved detection performance in environments with speckle noise, which is commonly found in adverse weather conditions.

2. Materials and Methods

2.1. Object Detection

Object detection models can be broadly classified into one-stage and two-stage detectors. One-stage detectors, which jointly learn to localize and classify objects using CNNs, are generally faster but less accurate than two-stage detectors. This makes them ideal for real-time object detection applications [17,18]. Representative one-stage detectors include SSD [19] and YOLO [20]. YOLO, which is the most commonly used one-stage detector, excels at extracting and detecting object and background features, as it learns from individual objects as well as the surrounding information and entire image domain.

The image input to YOLO is partitioned into a grid, $S \times S$, to extract object features through a convolutional layer and generate predicted tensors through a fully connected layer. Subsequently, for each partitioned grid cell, localization and classification are performed simultaneously to produce B candidate bounding boxes, along with the corresponding confidence score (CS) for each bounding box. Each bounding box contains information of $(x,\ y,\ \omega,\ h,\ CS)$, where $(x,\ y)$ refers to the coordinates of the center point of the bounding box normalized to each grid cell, and $(\omega,\ h)$ refers to the width and height of the bounding box. CS reflects the probability that the bounding box contains an object, i.e., the accuracy of the predicted box, and is defined as follows:

$$CS = P_{obj} \times IoU(pred,\ true) \quad (1)$$

Here, P_{obj} refers to the probability that the bounding box contains the object, with a value of 1 if the grid cell correctly contains the object and 0 otherwise. $IoU(pred,\ true)$ refers to the Intersection over Union (IoU) between the ground truth and predicted box, which is the width of the overlapping area and indicates how accurately the bounding box predicts the geometric information of the object. Moreover, the grid cell is expressed as a conditional probability of belonging to one of the C object classes within the bounding box, as presented below.

$$P_{class} = \Pr(class_i|object) \quad (2)$$

The class-specific confidence score (CCS), which represents the probability of an object being contained within each bounding box and the probability that the detected object matches the ground truth, is calculated as follows:

$$CCS = CS \times P_{Class} \quad (3)$$

The bounding box with the maximum CCS among the predicted B bounding boxes is chosen as the final bounding box for the target object.

2.2. Proposed Multi-Sensor-Based Detection Model

We propose a robust multi-sensor-based object detection model to prevent potential safety accidents caused by pedestrian detection failures in low-light or unexpected external noise environments. The proposed model performs object detection on RGB images using YOLO and simultaneously preprocesses the depth map generated by LiDAR using the proposed Gaussian–Sobel filter to increase the clarity of object contours; subsequently, this model performs object detection through a separate YOLO. At this time, to sensor-fuse the 2D image of the camera and the 3D PCD of LiDAR in parallel, a registration process is required to unify the type of image used for learning individual YOLO. Thereafter, NMS is applied to the individual detection results from the two sensors to determine the final object, as depicted in the block diagram presented in Figure 1. The proposed model is summarized by Algorithm 1.

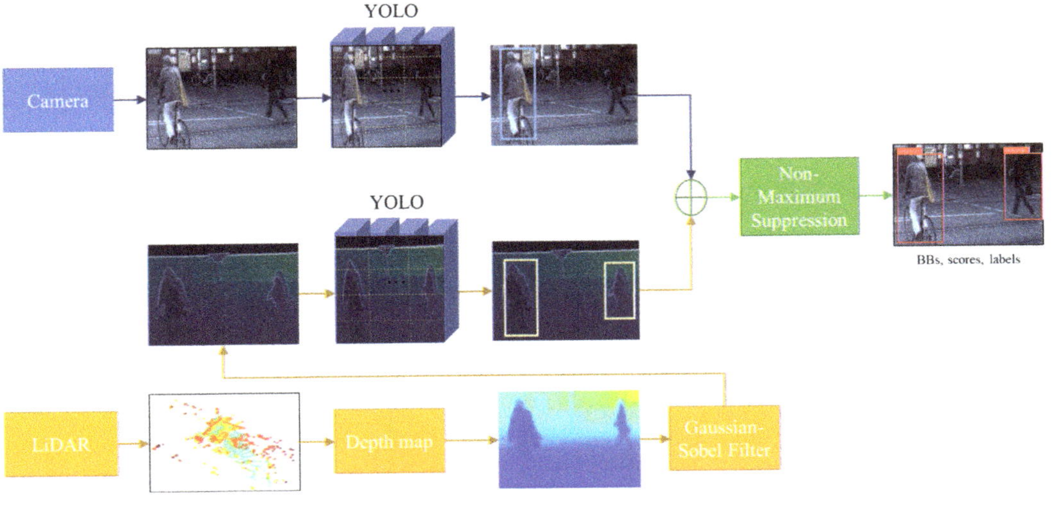

Figure 1. Block diagram of the proposed multi-sensor-based detection model.

Algorithm 1 Image Processing and Object Detection

```
Input: List of images (camera images and LiDAR Depth Maps)
Output: Performance metrics for each image
for each image in image_list do
  if image is camera_image then
    bboxes←YOLO(image)
    filtered_bboxes←NMS(bboxes, threshold)
else
    filtered_depth_map←GAUSSIAN_SOBEL_FILTER(image)
    bboxes←YOLO(filtered_depth_map)
    filtered_bboxes←NMS(bboxes, threshold)
  end if
  performance←EVALUATE_PERFORMANCE(filtered_bboxes)
  PRINT performance
end for
```

2.2.1. Creating a Depth Map for Image Registration

To effectively fuse data from heterogeneous sensors like cameras and LiDAR for object detection, it is essential to align the dimensions of 2D images and 3D PCD. This is typically achieved by transforming 3D PCD into a depth map. Furthermore, 3D PCD represent a set of 3D coordinate points, generated when laser signals emitted from LiDAR bounce off surrounding objects. To apply this transformation to a 2D setting, a calibration process is necessary to reconcile the disparate viewpoints of the two sensors beyond dimension transformation. This allows the $(x,\ y,\ z)$ coordinate of the PCD to be mapped to the $(u,\ v)$ coordinate of a 2D image, as expressed in the following equation [21]:

$$\begin{pmatrix} u \\ v \\ 1 \end{pmatrix} = \begin{pmatrix} f_u & 0 & u_0 \\ 0 & f_v & v_0 \\ 0 & 0 & 1 \end{pmatrix} \begin{pmatrix} R & t \\ 0 & 1 \end{pmatrix} \begin{pmatrix} x \\ y \\ z \\ 1 \end{pmatrix} \begin{pmatrix} x \\ y \\ z \\ 1 \end{pmatrix} \tag{4}$$

Here, f_u and f_v are the eigenvalues of the camera, representing the focal lengths in the horizontal and vertical directions, respectively; u_0 and v_0 denote the principal points of the 2D image; and R and t denote the rotational transformation and parallel translation matrices, respectively. They are computed using singular value decomposition as described in [22]. Consequently, a depth map, aligning with the camera's viewpoint, is created and employed for training YOLO. Figure 2 depicts the process of creating a depth map by aligning a LiDAR-acquired PCD point with a camera. Figure 2a presents an image acquired by an RGB camera, while Figure 2b displays a point map projected onto the image by aligning the 2D-transformed PCD with the camera. Moreover, Figure 2c illustrates the resulting depth map.

(**a**)

Figure 2. *Cont.*

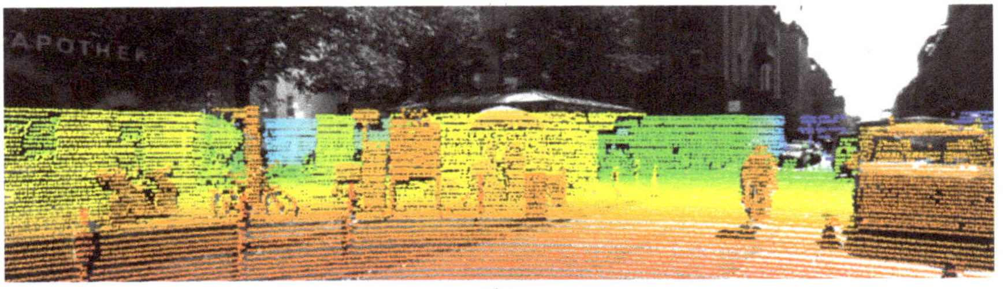

(b)

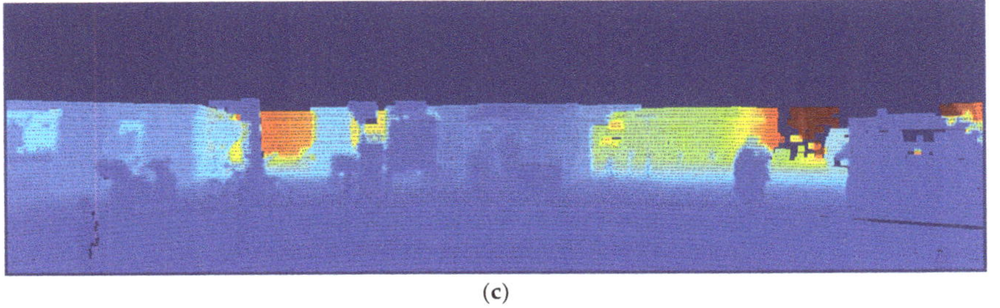

(c)

Figure 2. Process for generating a depth map for image registration: (**a**) RGB image; (**b**) PCD projected on RGB image; (**c**) depth map.

2.2.2. Preprocessing with the Gaussian–Sobel Filter

Object detection with RGB cameras suffers from significant performance degradation in night-time environments compared with daytime environments. To address this limitation, we propose a method that leverages depth maps obtained from LiDAR to enhance pedestrian detection in low-light conditions. Given that the depth map is derived from low-resolution PCD, object contours exhibit reduced sharpness. Consequently, an enhancement process is necessary, and a sharpening filter is additionally applied to address this issue. First, a Gaussian filter is applied to suppress the variance of depth information caused by the different sampling intervals of PCD depending on the LiDAR channel. Subsequently, a Sobel filter is utilized to accentuate the line edges of objects whose boundaries with the background become blurred owing to Gaussian blurring, thereby facilitating feature extraction of objects. By doing so, the model most effectively preserves pixel values of the depth information while further accentuating object contours.

A Gaussian filter generates a natural blurring effect on images by using a filter mask created by approximating a Gaussian distribution [23,24]. A 2D Gaussian distribution function with a mean of (0, 0) and standard deviations of σ_α and σ_β along the α and β axes, respectively, is defined as follows:

$$\Lambda_{\sigma_\alpha\sigma_\beta}(\alpha,\beta) = \frac{1}{2\pi\sigma_\alpha\sigma_\beta}e^{-\left(\frac{\alpha^2}{2\sigma_\alpha{}^2}+\frac{\beta^2}{2\sigma_\beta{}^2}\right)} \quad (5)$$

The depth map, L_d, can be expressed as follows after undergoing preprocessing by a Gaussian filter:

$$L_{d,g} = \Lambda_{\sigma_\alpha\sigma_\beta}(\alpha,\beta) * L_d \quad (6)$$

The Gaussian function peaks at $\Lambda_{\sigma_\alpha\sigma_\beta}(0,0)$ and decreases as the distance from the center increases. Furthermore, in the filter mask, pixels near the filtering target receive higher weights, whereas those further away receive lower weights, which helps mitigate the effects of noise pixels in the depth map.

Thus, the Gaussian filter suppresses the effects of noise in the depth map and blurs the overall image, resulting in a decrease in contrast. To compensate for this drawback, a Sobel filter is directly applied to preserve as much depth information of the object as possible while still accentuating object contours. The Sobel filter uses two 3 × 3 kernels to detect edges: one is for finding changes along the horizontal direction, while the other is for finding changes in the vertical direction. Edges are points where the instantaneous rate of change of a function is large. Therefore, if a value exceeding a certain threshold is derived by calculating the differential value at each pixel of the image, the corresponding point is determined as an edge [25,26]. The two kernels are convolved with the original image to approximate the rate of change; if we define $L^x_{d,g-s}$and $L^y_{d,g-s}$ as two images containing approximations of the horizontal and vertical derivatives, respectively, then the corresponding calculation is as follows:

$$L^x_{d,g-s} = \begin{bmatrix} 1 & 0 & -1 \\ 2 & 0 & -2 \\ 1 & 0 & -1 \end{bmatrix} * L_{d,g}\ ,\ L^y_{d,g-s} = \begin{bmatrix} -1 & -2 & -1 \\ 0 & 0 & 0 \\ 1 & 2 & 1 \end{bmatrix} * L_{d,g} \tag{7}$$

Here, the x-coordinate is defined as increasing to the right, and the y-coordinate is defined as increasing downward. Furthermore, at each pixel in the image, $L^x_{d,g-s}$ and $L^y_{d,g-s}$ are combined to determine the size, $L_{d,g-s} = \sqrt{\left(L^x_{d,g-s}\right)^2 + \left(L^y_{d,g-s}\right)^2}$, and direction, $\Theta = \text{atan}\left(\frac{L^y_{d,g-s}}{L^x_{d,g-s}}\right)$, of the gradient. Consequently, the variance and noise of LiDAR PCD's depth information are suppressed, and the object is simultaneously separated from the background. This can be verified by referring to Figure 3, which shows an example of the sequential application of two filters to the depth map.

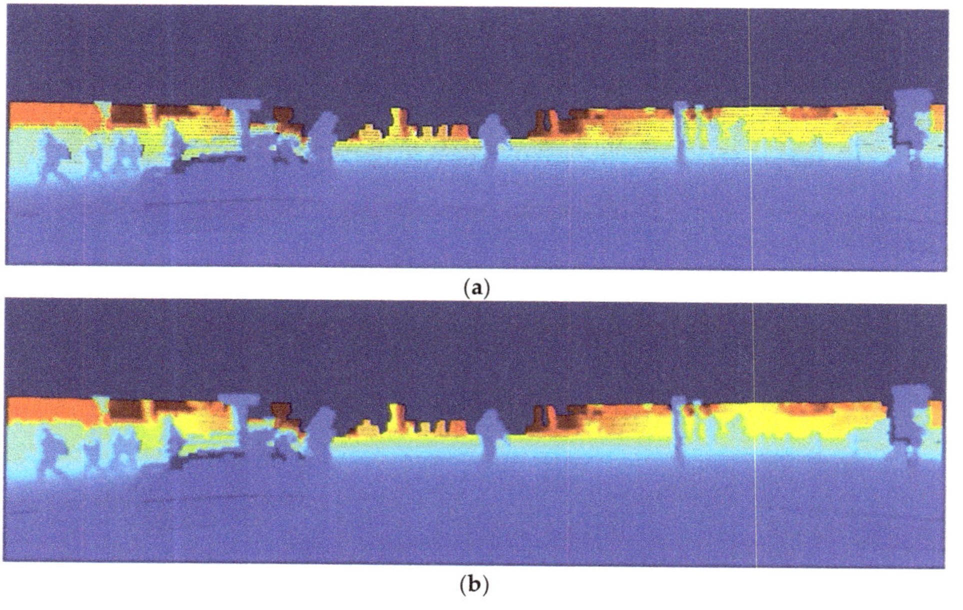

Figure 3. *Cont.*

(c)

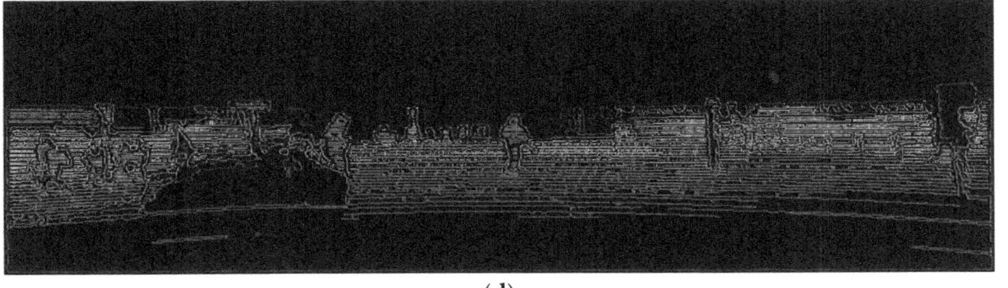

(d)

Figure 3. Preprocessing of depth maps using the Gaussian–Sobel filter: (**a**) depth map; (**b**) depth map after Gaussian filtering; (**c**) depth map after Gaussian–Sobel filtering; (**d**) depth map after Canny edge filtering.

As mentioned earlier, Figure 3a illustrates the variance of depth information in PCD, which varies with the data generation interval. To suppress this, we applied a Gaussian filter (Figure 3b), which suppresses the effects of noise and smooths out the contour information of the object. Moreover, we applied an additional Sobel filter, resulting in a clearer delineation of object contours, as shown in Figure 3c.

Notably, the use of Gaussian blurring and Sobel operators is also a process included in the Canny edge filter. Furthermore, the reasons for the mandatory use of the Gaussian–Sobel filter are as follows: after Gaussian blurring and Sobel masking, the Canny edge filter [27] further applies NMS and hysteresis edge tracking. However, the noise from the LiDAR channel is mistakenly enhanced along with the edges, resulting in a failure to properly segment objects from the background, thereby leading to inaccurate object detection. This phenomenon is illustrated in Figure 3d. Thus, by applying a Gaussian filter and Sobel operator, it is possible to suppress the effects of noise that deteriorates the object detection performance of the depth map; moreover, more accurate object detection is possible while maintaining the original pixel values for object contours.

2.2.3. Object Estimation Using NMS

We obtain object detection results from a YOLO model applied to RGB images and another YOLO model applied to preprocessed depth maps. By fusing these results, we achieve optimized object estimation. The bounding boxes of objects detected through two independent YOLOs have CS, and the degree of overlap of each bounding box is determined based on the IoU. To fuse the results of individually performed object detection and ensure higher object estimation performance, it is necessary to select bounding boxes with high confidence by recognizing overlapping bounding boxes and their respective CS values; NMS is performed for this purpose [28]. If the IoU value between two bounding boxes is greater than a specified threshold, it is determined that the bounding boxes have detected the same object, and the bounding box with a higher CS is selected. This process is repeated until there are no more remaining bounding boxes (Figure 4), thereby selecting the bounding box with the highest *CS* from the remaining bounding boxes each time.

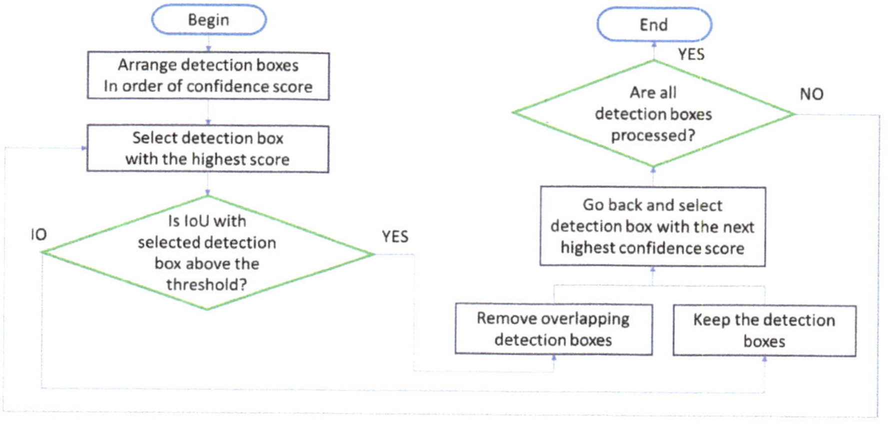

Figure 4. Flowchart for non-maximum suppression (NMS).

3. Experimental Results

To evaluate the performance of the proposed method, we generated RGB images, images with various filters applied to the LiDAR depth map, and simulated images considering the presence of noise in LiDAR. Based on this, we aimed to verify the superiority of the proposed object detection model through comparisons with existing models that utilize RGB images or LiDAR depth maps for object detection, as well as models that fuse preprocessed depth maps.

3.1. Experimental Environment

Our proposed model is designed to reliably detect pedestrians under challenging conditions such as low light and adverse weather. It employs YOLOv4 for object detection and is implemented on an NVIDIA RTX 3060 (Santa Clara, CA, USA) and an Intel Core i7-12700 CPU (Santa Clara, CA, USA). YOLOv4 [15] is compatible with a range of deep learning frameworks, including TensorFlow and PyTorch, and offers versatility and scalability in various environments. This is particularly relevant in domains such as autonomous driving and CCTV monitoring, where real-time pedestrian detection is of paramount importance. The selection of YOLOv4 as our detection model was driven by the necessity to detect pedestrians of varying shapes in complex environments. Furthermore, there is a potential for extending its capabilities to embedded systems in the future. For the test evaluation, we used the KITTI Open Dataset [29], which consists of 1467 RGB camera images and LiDAR PCD acquired at the same time and location. Of these images, 1200 (approximately 80%) were employed for model training, with the remaining 267 being used for testing. All images were resized to 1242 pixels × 375 pixels, ensuring consistency between the RGB and depth maps. Moreover, the dataset was restricted to pedestrian-class instances.

The proposed object detection model, trained on various environments of autonomous driving scenarios, was evaluated using AP as a performance metric. AP evaluates the model performance by calculating the area under the precision–recall curve. Precision is defined as TP/(TP + FP), representing the proportion of correctly detected instances among all detected instances, while recall is defined as TP/(TP + FN), representing the proportion of correctly detected instances among those that need to be detected. Here, TP, FP, and FN stand for True Positive, False Positive, and False Negative, respectively [30].

3.2. Performance Evaluation of Object Detection under Varying Brightness Levels

To evaluate the performance of our proposed model, we compared it with state-of-the-art models that fuse RGB cameras and LiDAR data and share similar architectures. The comparative results are tabulated in Table 1. The experimental results indicate that the RGB-LiDAR fusion models, which represent the most common architecture, and the novel model proposed in this study achieved high performance levels during daytime scenarios, with the brightness set to 100%. No significant performance gap was observed between the RGB-LiDAR fusion models and the proposed model. However, the performance of both models gradually deteriorated as the lighting conditions worsened, primarily due to the RGB camera's hypersensitivity to low light. When the image brightness was reduced to 40% of its original level, the proposed model demonstrated a slightly higher average AP compared to the baseline model, with a performance gap of approximately 1.5%. We also compared our proposed model with models that apply preprocessing filters to depth maps with a similar architecture, such as Maragos and Pessoa [12], Deng [13], and Ali and Clausi [14]. It was found that all models perform well in daylight and do not seem to differ significantly, while in darkness, our proposed model shows 1–2% higher APs and a relatively small improvement in detection accuracy.

Table 1. Comparison of pedestrian detection performance based on average precision (AP) [%] between the proposed model and similar models under varying brightness levels.

Model	Filter Used for Depth Map Preprocessing	Brightness Level		
		100%	70%	40%
Depth Map	-	83.49	83.49	83.49
Depth Map + RGB	-	91.99	91.32	85.60
Maragos and Pessoa [12]	Morphology dilation	91.94	91.00	86.05
Deng [13]	Unsharp Mask	92.13	90.72	85.76
Ali and Clausi [14]	Canny Edge	92.43	91.09	85.08
Proposed model	Gaussian–Sobel	92.07	91.49	87.03

Figure 5 shows an example of pedestrian detection results in daylight, i.e., when the brightness level is 100%. In the figure, the white bounding box indicates the ground truth, the blue box indicates a successful pedestrian detection, and the thick yellow box indicates a missed detection. Figure 5a shows the result of detecting pedestrians using only the depth map obtained from LiDAR, while Figure 5b shows the result of fusing LiDAR with an RGB camera to detect pedestrians. Figure 5c–f show the detection results of the Maragos and Pessoa [12] model, which sequentially accentuates the contours of objects with the Dilation filter; the Deng [13] model, which emphasizes the density of objects with the unsharp mask filter; the Ali and Clausi [14] model applied with the Canny edge filter; and the proposed model, respectively. As can be seen, when the illumination was sufficiently high, both existing models with similar architectures detected most of the objects except for one or two, and no substantial difference was observed in their performance. However, a closer inspection reveals that the proposed model distinctly identifies a pedestrian located centrally in the scene, which is overlooked by all other methods. The proposed model's superior reconstruction quality facilitates more accurate detection.

To evaluate the effectiveness of the proposed model for detecting objects in low-light environments, we conducted a pedestrian detection experiment by artificially manipulating the image to decrease its brightness to <40% of the brightness of the original image; an example of the results is shown in Figure 6. The false alarms represented by the thick pink boxes shown in Figure 6a,b were generated by models that did not apply a sharpening filter to the depth map, which can be attributed to the difficulty of distinguishing inaccurate pedestrian shapes with the depth map alone, or to the fact that the camera's role considerably decreases with decreasing illumination. However, Figure 6c–f show that by sharpening the objects with a preprocessing filter on the depth map, the false alarm issue disappears altogether. Furthermore, Figure 6c,e show that the preprocessing of the depth map actually blurs the contours of some objects, resulting in non-detections. This demonstrates that the use of a sharpening filter can positively affect the shape of certain objects, while adversely affecting others, leading to blurred object contours. Nevertheless, the proposed model solved the problems of missed detection and false alarms of other models by making object contours clearer compared with other models and stably detected all pedestrians.

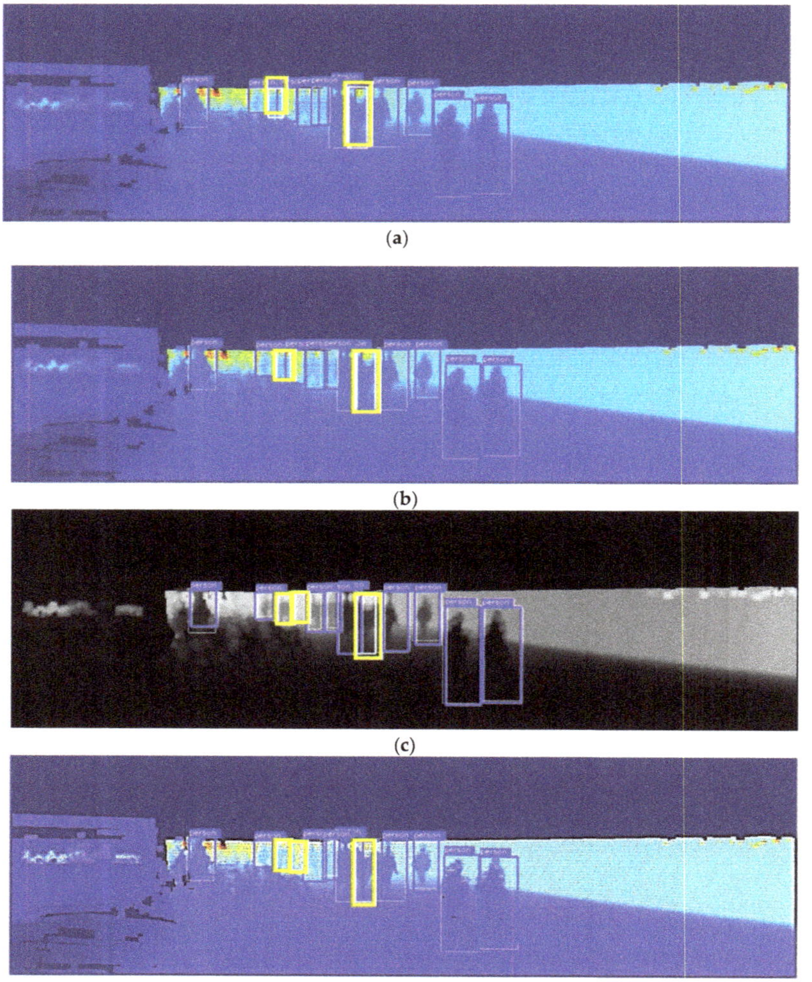

(**a**) (**b**) (**c**) (**d**)

Figure 5. *Cont.*

(e)

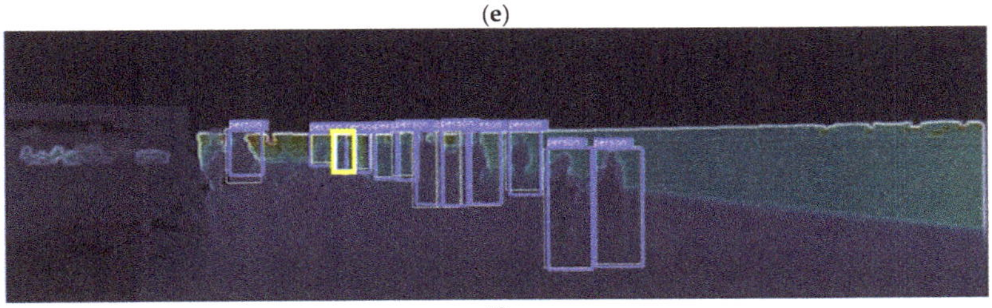

(f)

Figure 5. Comparison of pedestrian detection performance of the proposed model and similar models at 100% brightness: (**a**) depth map; (**b**) RGB + depth map; (**c**) Maragos and Pessoa [12]; (**d**) Deng [13]; (**e**) Ali and Clausi [14]; (**f**) proposed model.

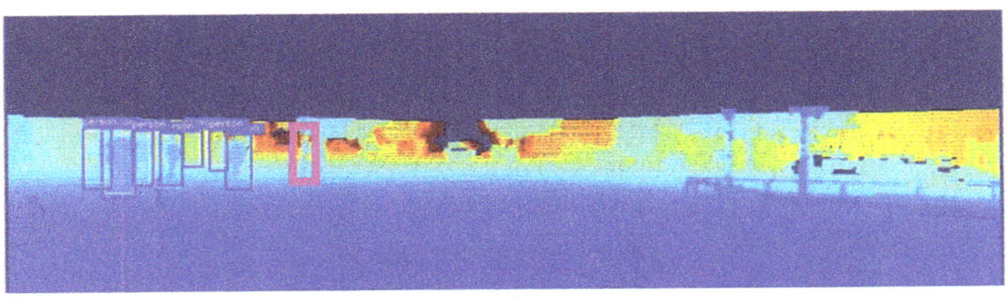

(a)

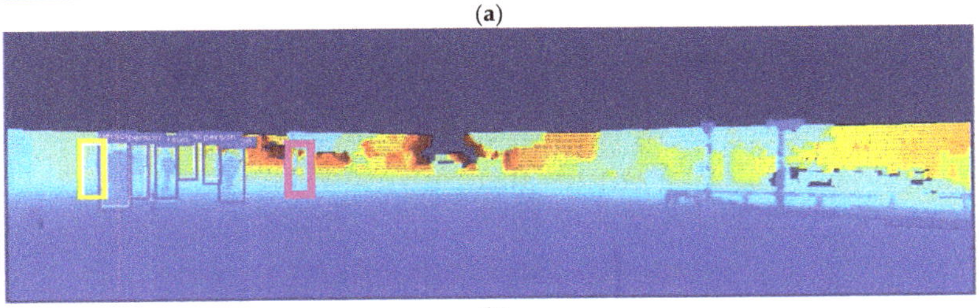

(b)

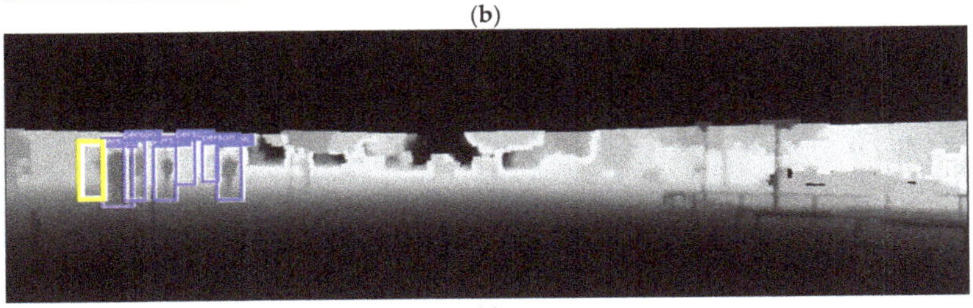

(c)

Figure 6. *Cont.*

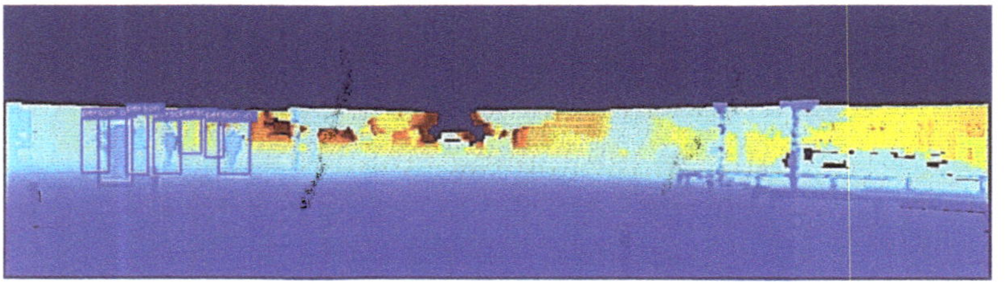

(**d**)

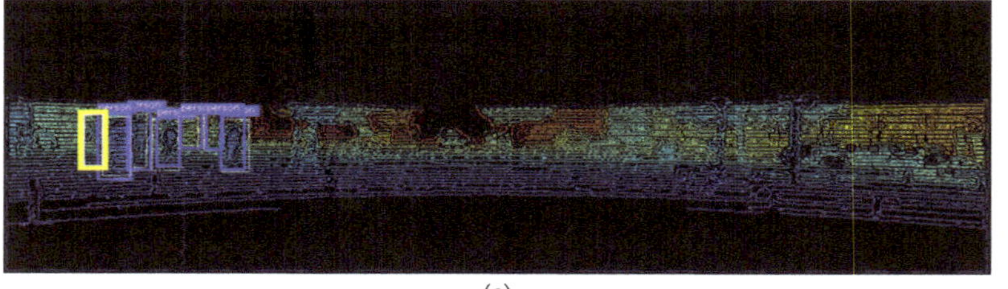

(**e**)

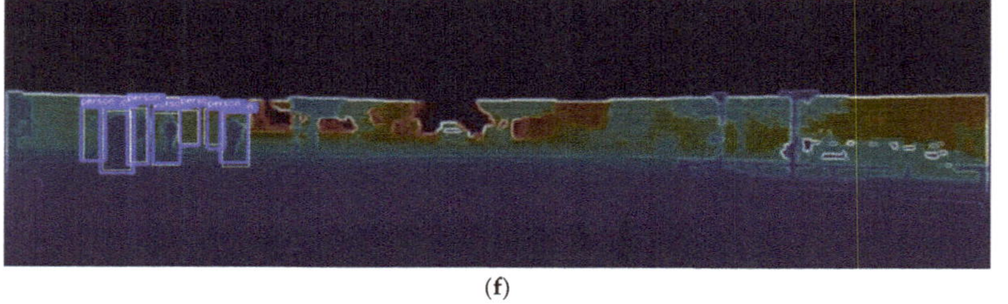

(**f**)

Figure 6. Comparison of pedestrian detection performance of the proposed model and similar models at 40% brightness level: (**a**) depth map; (**b**) RGB + depth map; (**c**) Maragos and Pessoa [12]; (**d**) Deng [13]; (**e**) Ali and Clausi [14]; (**f**) proposed model.

3.3. Evaluation of Detection Performance under Varying Noises

As shown by the aforementioned experimental results, the performance of the proposed model is slightly better than the existing models in various low-light environments. Owing to its robustness against external factors such as light reflection, shadows, and darker conditions, LiDAR is far more reliable in low-light environments than cameras are. However, as LiDAR data are generated in the form of point clouds, the model may be susceptible to adverse weather conditions such as snow, rain, and fog. Therefore, to evaluate the proposed model's coping ability when LiDAR is exposed to external environmental noise, additional experiments were carried out by adding Gaussian noise, a common type of noise encountered in real-world conditions, to the original images. To isolate the impact of low-light conditions on detection performance, we conducted experiments by varying the Gaussian noise variance in an environment with 40% ambient brightness. The pedestrian detection performance of our proposed model was compared with those of similar models. The results are summarized in Table 2. Our findings indicate that the object detection model relying solely on depth maps exhibited a substantial decline in detection performance as the noise levels increased, with its accuracy reaching as low as 50%. By integrating this model with camera data, this performance degradation was mitigated. Similarly, models that incorporated additional preprocessing on depth maps suffered from performance degradation due to increased noise. However, a camera mitigated this issue, ensuring that the detection performance remained at a reasonable 60–70% level. Compared with other models, our proposed model exhibited superior robustness to noise, maintaining approximately 80% of its performance even under increased noise. Existing models are deficient in noise effect suppression and solely focus on the contour information of all objects. Alternatively, our proposed model, which smoothens the scale of LiDAR by considering it to be noise and further overlaying the contour information of potentially occluded objects, demonstrates the most superior performance.

Table 2. Comparison of pedestrian detection performance based on AP [%] with the proposed model and similar models under noise variation in a 40% brightness environment.

Model	Filter Used for Depth Map Preprocessing	Noise Level		
		0%	0.2%	0.5%
Depth Map	-	83.49	69.84	51.26
Depth Map + RGB	-	85.60	80.34	75.37
Maragos and Pessoa [12]	Morphology dilation	86.05	74.88	65.55
Deng [13]	Unsharp Mask	85.76	77.87	68.75
Ali and Clausi [14]	Canny Edge	85.08	81.96	77.62
Proposed model	Gaussian–Sobel	87.03	86.29	84.72

Figure 7 shows the object detection results based on an image generated by assuming that Gaussian noise with a variance of 0.5% can be introduced through LiDAR. Figure 7a shows that in the case of detecting objects using only the depth map, most objects cannot be detected, and it completely fails to function as a detection model. In contrast, Figure 7b–f show a considerable decrease in the number of undetected objects by fusing RGB images and applying preprocessing to the depth map. In particular, the proposed model had a higher detection performance than other models and detected all objects. These results indicate that the proposed model exhibits strong noise robustness, demonstrating its reliability in various low-light night-time scenarios and noisy environments.

(**a**)

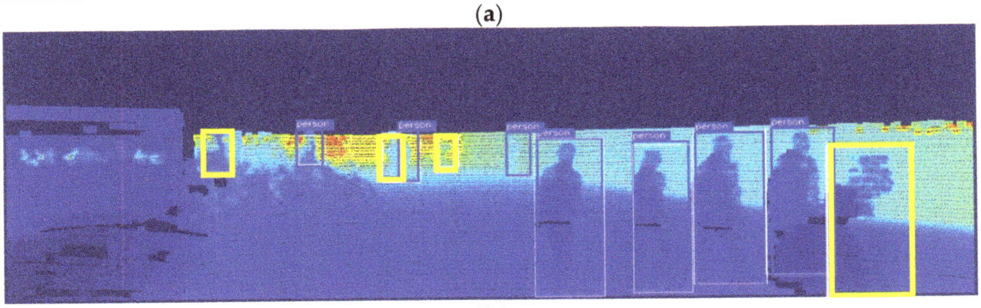

(**b**)

Figure 7. *Cont.*

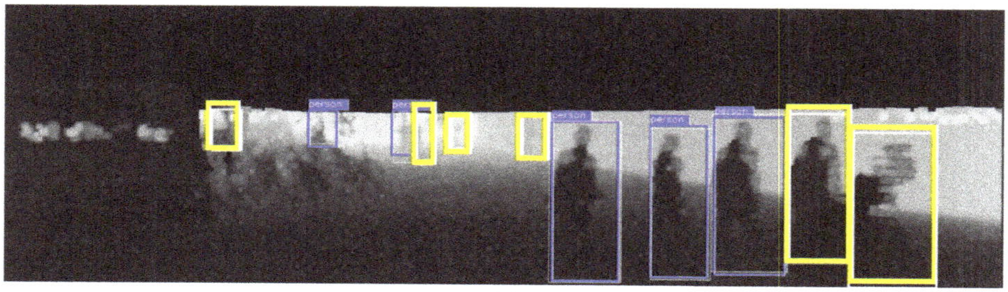

(**c**)

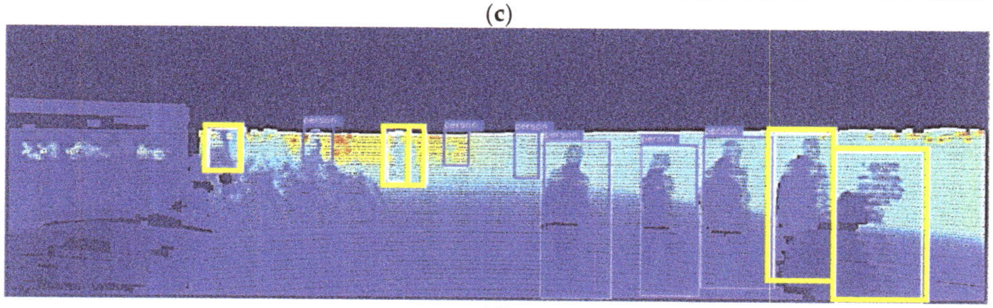

(**d**)

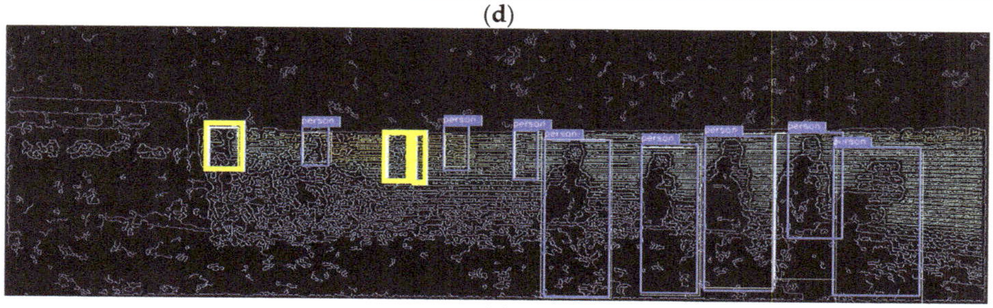

(**e**)

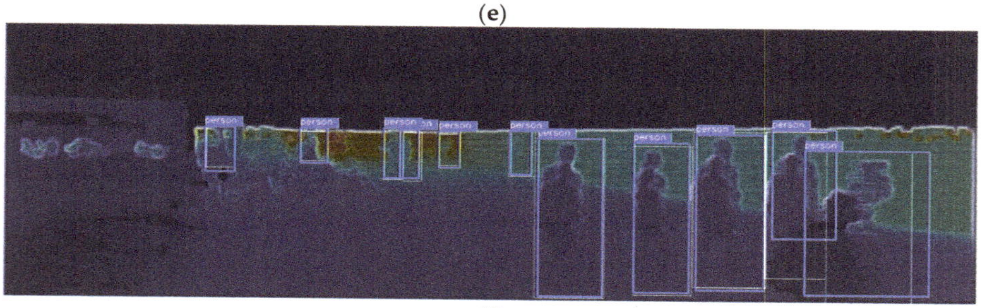

(**f**)

Figure 7. Comparison of the pedestrian detection performance of the proposed model and similar models at 40% brightness and 0.5% noise level: (**a**) depth map; (**b**) RGB + depth map; (**c**) Maragos and Pessoa [12]; (**d**) Deng [13]; (**e**) Ali and Clausi [14]; (**f**) proposed model.

4. Discussion

Despite advancements in autonomous vehicle technology and increasing utilization rates, object detection models that are reliant on cameras and LiDAR remain susceptible to causing pedestrian collision accidents under adverse conditions such as night-time or inclement weather. Therefore, this study proposes a new pedestrian detection model utilizing multiple sensors and fusion filters to improve pedestrian detection performance by compensating for the weaknesses of such detection models, thereby demonstrating an innovative method for enhancing object identification in low-light conditions. To effectively utilize depth maps created through LiDAR for object detection, we combined three key elements: a Gaussian blurring function to suppress the effects of noise, in combination with the Sobel operator to accentuate pedestrian features, and optimization of pedestrian detection estimation through the fusion of heterogeneous sensors, camera, and LiDAR. Through experimentation, we verified the potential of this combination to considerably improve pedestrian detection accuracy. Our results indicate that this approach outperforms traditional methods in terms of detection accuracy. In particular, this approach maintains a remarkable detection accuracy even in low-light and noisy environments where object features and contours are not clearly visible. Moreover, this approach demonstrates its efficiency by achieving at least a 7% improvement in AP compared with previously reported approaches. Future research will focus on overcoming these limitations and enhancing the robustness of the system. Furthermore, we aim to implement the proposed model on embedded systems and reduce its size for deployment on devices with limited computing resources, thereby facilitating practical applications. To address these limitations and explore future possibilities, we anticipate advancements in approaches for developing more robust low-light object detection models that can be deployed on a wider range of devices.

Author Contributions: M.J. and J.C. took part in the discussion of the work described in this paper. All authors have read and agreed to the published version of the manuscript.

Funding: This research was funded by a National Research Foundation of Korea (NRF) grant funded by the Korean government (MOE) (No. 2021R1I1A3055973) and the Soonchunhyang University Research Fund.

Institutional Review Board Statement: Not applicable.

Informed Consent Statement: Not applicable.

Data Availability Statement: The raw data supporting the conclusions of this article will be made available by the authors on request.

Acknowledgments: The authors thank the editor and anonymous reviewers for their helpful comments and valuable suggestions.

Conflicts of Interest: The authors declare no conflicts of interest.

References

1. Claussmann, L.; Revilloud, M.; Gruyer, D.; Glaser, S. A Review of Motion Planning for Highway Autonomous Driving. *IEEE Trans. Intell. Transp. Syst.* **2020**, *21*, 1826–1848. [CrossRef]
2. Bresson, G.; Alsayed, Z.; Yu, L.; Glaser, S. Simultaneous Localization and Mapping: A Survey of Current Trends in Autonomous Driving. *IEEE Trans. Intell. Veh.* **2017**, *2*, 194–220. [CrossRef]
3. Jhong, S.; Chen, Y.; Hsia, C.; Wang, Y.; Lai, C. Density-Aware and Semantic-Guided Fusion for 3-D Object Detection Using LiDAR-Camera Sensors. *IEEE Sens. J.* **2023**, *23*, 22051–22063. [CrossRef]
4. Cheng, L.; He, Y.; Mao, Y.; Liu, Z.; Dang, X.; Dong, Y.; Wu, L. Personnel Detection in Dark Aquatic Environments Based on Infrared Thermal Imaging Technology and an Improved YOLOv5s Model. *Sensors* **2024**, *24*, 3321. [CrossRef] [PubMed]
5. Hsu, W.; Yang, P. Pedestrian Detection Using Multi-Scale Structure-Enhanced Super-Resolution. *IEEE Trans. Intell. Transp. Syst.* **2023**, *24*, 12312–12322. [CrossRef]
6. Zhang, T.; Ye, Q.; Zhang, B.; Liu, J.; Zhang, X.; Tian, Q. Feature Calibration Network for Occluded Pedestrian Detection. *IEEE Trans. Intell. Transp. Syst.* **2022**, *23*, 4151–4163. [CrossRef]
7. Mushtaq, Z.; Nasti, S.; Verma, C.; Raboaca, M.; Kumar, N.; Nasti, S. Super Resolution for Noisy Images Using Convolutional Neural Networks. *Mathematics* **2022**, *10*, 777. [CrossRef]
8. Xu, X.; Wang, S.; Wang, Z.; Zhang, X.; Hu, R. Exploring Image Enhancement for Salient Object Detection in Low Light Images. *ACM Trans. Multimed. Comput. Commun. Appl.* **2021**, *17*, 1–19. [CrossRef]
9. Gilroy, S.; Jones, E.; Glavin, M. Overcoming Occlusion in the Automotive Environment—A Review. *IEEE Trans. Intell. Transp. Syst.* **2020**, *22*, 23–35. [CrossRef]

10. Lin, T.; Tan, D.; Tang, H.; Chien, S.; Chang, F.; Chen, Y.; Cheng, W. Pedestrian Detection from Lidar Data via Cooperative Deep and Hand-Crafted Features. In Proceedings of the IEEE International Conference on Image Processing, Athens, Greece, 7–10 October 2018; pp. 1922–1926.
11. Qi, C.; Liu, W.; Wu, C.; Su, H.; Guibas, L. Frustum PointNets for 3D Object Detection from RGB-D Data. In Proceedings of the IEEE Conference on Computer Vision and Pattern Recognition, Salt Lake City, UT, USA, 18–23 June 2018; pp. 918–927.
12. Maragos, P. Morphological Filtering for Image Enhancement and Feature Detection. In *The Image and Video Processing Handbook*, 2nd ed.; Bovik, A.C., Ed.; Elsevier Academic Press: Cambridge, MA, USA, 2005; pp. 135–156.
13. Deng, G. A Generalized Unsharp Masking Algorithm. *IEEE Trans. Image Process.* **2011**, *20*, 1249–1261. [CrossRef]
14. Ali, M.; Clausi, D. Using the Canny Edge Detector for Feature Extraction and Enhancement of Remote Sensing Images. In Proceedings of the IEEE International Geoscience and Remote Sensing Symposium, Sydney, Australia, 9–13 July 2001; pp. 2298–2300.
15. Bochkovskiy, A.; Chien, W.; Hong, L. Yolov4: Optimal Speed and Accuracy of Object Detection. *arXiv* **2020**, arXiv:2004.10934.
16. Ravpreet, K.; Sarbjeet, S. A Comprehensive Review of Object Detection with Deep Learning. *Digit. Signal Process.* **2023**, *132*, 203812.
17. Pham, M.; Courtrai, L.; Friguet, C.; Lefèvre, S.; Baussard, A. YOLO-Fine: One-Stage Detector of Small Objects Under Various Backgrounds in Remote Sensing Images. *Remote Sens.* **2020**, *12*, 2501. [CrossRef]
18. Chen, K.; Li, J.; Lin, W.; See, J.; Wang, J.; Duan, L.; Chen, Z.; He, C.; Zou, J. Towards Accurate One-Stage Object Detection with AP-Loss. In Proceedings of the IEEE/CVF Conference on Computer Vision and Pattern Recognition, Long Beach, CA, USA, 15–20 June 2019; pp. 5114–5122.
19. Liu, W.; Anguelov, D.; Erhan, D.; Szegedy, C.; Reed, S.; Fu, C.; Berg, A. SSD: Single Shot Multibox Detector. In Proceedings of the European Conference on Computer Vision, Amsterdam, The Netherlands, 8–16 October 2016; pp. 21–37.
20. Redmon, J.; Divvala, S.; Girshick, R.; Farhadi, A. You Only Look Once: Unified, Real-Time Object Detection. In Proceedings of the IEEE Conference on Computer Vision and Pattern Recognition, Las Vegas, NV, USA, 27–30 June 2016; pp. 779–788.
21. Cai, H.; Pang, W.; Chen, X.; Wang, Y.; Liang, H. A Novel Calibration Board and Experiments for 3D LiDAR and Camera Calibration. *Sensors* **2020**, *20*, 1130. [CrossRef] [PubMed]
22. Xie, X.; Wang, C.; Li, M. A Fragile Watermark Scheme for Image Recovery Based on Singular Value Decomposition, Edge Detection and Median Filter. *Appl. Sci.* **2019**, *9*, 3020. [CrossRef]
23. Tang, D.; Xu, Y.; Liu, X. Application of an Improved Laplacian-of-Gaussian Filter for Bearing Fault Signal Enhancement of Motors. *Machines* **2024**, *12*, 389. [CrossRef]
24. Popkin, T.; Cavallaro, A.; Hands, D. Accurate and Efficient Method for Smoothly Space-Variant Gaussian Blurring. *IEEE Trans. Image Process.* **2010**, *19*, 1362–1370. [CrossRef]
25. Ma, Y.; Ma, H.; Chu, P. Demonstration of Quantum Image Edge Extraction Enhancement through Improved Sobel Operator. *IEEE Access* **2020**, *8*, 210277–210285. [CrossRef]
26. Kanopoulos, N.; Vasanthavada, N.; Baker, R. Design of an Image Edge Detection Filter Using the Sobel Operator. *IEEE J. Solid-State Circuits* **1988**, *23*, 358–367. [CrossRef]
27. Pawar, K.; Nalbalwar, S. Distributed Canny Edge Detection Algorithm Using Morphological Filter. In Proceedings of the IEEE International Conference on Recent Trends in Electronics, Information & Communication Technology, Bangalore, India, 20–21 May 2016; pp. 1523–1527.
28. Zaghari, N.; Fathy, M.; Jameii, S.; Shahverdy, M. The improvement in obstacle detection in autonomous vehicles using YOLO non-maximum suppression fuzzy algorithm. *J. Supercomput.* **2021**, *77*, 13421–13446. [CrossRef]
29. Geiger, A.; Lenz, P.; Urtasun, R. Are we ready for autonomous driving? The kitti vision benchmark suite. In Proceedings of the IEEE Conference on Computer Vision and Pattern Recognition, Providence, RI, USA, 16–21 June 2012; pp. 3354–3361.
30. Behl, A.; Mohapatra, P.; Jawahar, C.; Kumar, M. Optimizing Average Precision Using Weakly Supervised Data. *IEEE Trans. Pattern Anal. Mach. Intell.* **2015**, *37*, 2545–2557. [CrossRef] [PubMed]

applied sciences

MDPI

Article

Multi-Scale Feature Fusion and Context-Enhanced Spatial Sparse Convolution Single-Shot Detector for Unmanned Aerial Vehicle Image Object Detection

Guimei Qi [1,2], Zhihong Yu [2,*] and Jian Song [2]

[1] College of Computer Science and Technology, Inner Mongolia Normal University, Hohhot 010022, China; ciecqgm@imnu.edu.cn

[2] College of Mechanical and Electrical Engineering, Inner Mongolia Agricultural University, Hohhot 010010, China; songjian9703@emails.imau.edu.cn

* Correspondence: yzhyq@imau.edu.cn

Abstract: Accurate and efficient object detection in UAV images is a challenging task due to the diversity of target scales and the massive number of small targets. This study investigates the enhancement in the detection head using sparse convolution, demonstrating its effectiveness in achieving an optimal balance between accuracy and efficiency. Nevertheless, the sparse convolution method encounters challenges related to the inadequate incorporation of global contextual information and exhibits network inflexibility attributable to its fixed mask ratios. To address the above issues, the MFFCESSC-SSD, a novel single-shot detector (SSD) with multi-scale feature fusion and context-enhanced spatial sparse convolution, is proposed in this paper. First, a global context-enhanced group normalization (CE-GN) layer is developed to address the issue of information loss resulting from the convolution process applied exclusively to the masked region. Subsequently, a dynamic masking strategy is designed to determine the optimal mask ratios, thereby ensuring compact foreground coverage that enhances both accuracy and efficiency. Experiments on two datasets (i.e., VisDrone and ARH2000; the latter dataset was created by the researchers) demonstrate that the MFFCESSC-SSD remarkably outperforms the performance of the SSD and numerous conventional object detection algorithms in terms of accuracy and efficiency.

Keywords: UAV image object detection; SSD; multi-scale feature fusion; context-enhanced spatial sparse convolution

1. Introduction

Vehicles (UAVs), as a novel and prominent sensing platform, have become increasingly significant in a variety of fields due to their ability to capture high-resolution images. Object detection based on UAV images, which aims to detect object instances of predefined categories, has become a popular research topic. However, unlike objects in natural scene datasets such as COCO [1], the objects in UAV images are characterized by large numbers with complex backgrounds. Most of these objects are small in size (Figure 1), which significantly increases the difficulty of object detection in UAV images. Meanwhile, UAV hardware is often resource-constrained, leading to an urgent need for lightweight models for fast inference. Accurate and fast object detection is a typical problem encountered in the application of UAV images.

https://doi.org/10.3390/app15020924

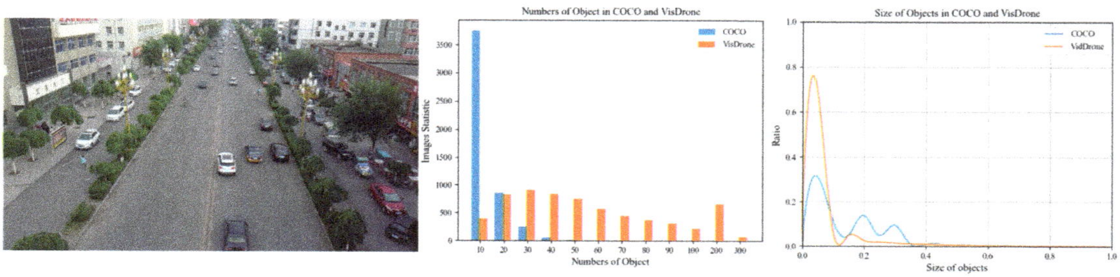

Figure 1. A visualization of objects in a sample image from VisDrone2019 (classical UAV dataset), and a comparison of objects in the UAV image and COCO datasets. The number of objects in each VisDrone2019 sample is uniformly distributed between 10 and 300, while the number of objects in each COCO sample is mostly less than 20. The percentage of small objects (with a ratio of 0.05 to the entire background) in VisDrone is up to 72.45%.

A single-shot multibox detector (SSD), which combines the regression idea of YOLO and the anchor mechanism of Faster R-CNN to perform regression in multi-scale feature maps, demonstrated promising performance in multi-scale object detection tasks, both in terms of efficiency and accuracy [2]. For multi-scale feature maps, the shallow feature is used to detect small objects, while the deep layer feature is utilized to detect large objects. This strategy improves detection accuracy because small objects retain sufficient spatial information in shallow layers, while large objects are recognized and located by deep layer features with large receptive fields. In conclusion, the SSD showed potential in multi-scale object detection tasks [3].

However, each layer of multi-scale feature maps is individually responsible for detecting its corresponding size object, which leads to the poor performance of the SSD in detecting dense small objects. One reason is that the features of small objects generated by Conv4-3 lack sufficient semantic information. Additionally, small objects also heavily rely on contextual information. Aiming to address the aforementioned problems, various strategies and methods have been explored. A feature fusion single-shot multibox detector (FSSD) extracts features from different layers and concatenates them together, followed by down-sampling some blocks to generate a feature pyramid [4]. Furthermore, attention mechanisms are introduced into the SSD structure. For instance, an attention and feature fusion SSD (AF-SSD) uses the attention mechanism to enhance or suppress the channel dimension of the feature [5]. Similarly, an improved SSD network employs a hybrid attention unit and focal loss to improve the imbalance of positive and negative samples [6]. In UAV imagery, the target typically constitutes a minor fraction of the foreground. Extensive computation of the entire foreground and background by the SSD using dense detection heads significantly impacts the efficiency and real-time capabilities of the detection process.

Spatial sparse convolution (SSC) [7] only performs convolution on sparse regions via a mask, which represents a promising alternative for increasing the speed of training and testing. SSC was first applied for object detection tasks based on 3D data, such as LiDAR or RGB-D data [8]. Detection heads must handle numerous bounding boxes, especially in one-stage methods, which are generated by general 2D image detection algorithms. Several approaches have recently used SSC as a detection head to save computation costs and increase efficiency. For instance, QueryDet based on cascade sparse query initially predicts the coarse location of objects on low-resolution feature maps and then guides the high-resolution feature maps to infer the accuracy of the location based on this coarse location [9]. A fine-grained dynamic router provides three kinds of different scale features for each detection head and adopts SSC to release the efficiency of fine dynamic routers [10].

The above study established a meaningful foreground by setting a fixed mask ratio, which reduces the number of parameters and the number of floating-point operations. However, a fixed mask ratio ignores the fact that the size of the foreground region varies with the flight altitude and viewing angle, reducing the flexibility of the network.

An SSD is leveraged in this paper, and a novel model, namely a multi-scale feature fusion and context-enhanced spatial sparse convolution SSD (MFFCESSC-SSD), is proposed to facilitate effective object detection in UAV images. First, a multi-scale feature fusion (MFF) scheme is defined to project and concatenate features obtained from different scales followed by a batch normalization (BN). Self-attention units (SAUs) are used to capture the internal correlation between features and suppress the effect of background noise while obtaining global contextual information. Then, down-sampling and horizontal connection are applied to generate a new feature pyramid, which is then fed to the detection heads. Second, context-enhanced spatial sparse convolution (CESSC) is developed to optimize the detection heads and enhance meaningful areas within the foreground, which comprises context-enhanced group normalization (CE-GN) and an optimal mask estimation mechanism based on ground-truth labeling. MFF merges the local detailed and global semantic features to confirm the detection of small objects in UAV images, significantly improving detection accuracy. Meanwhile, CESSC aims to optimize the detection heads in different layers and integrate focal and context-enhanced information via limiting computations by only performing convolution on sparse regions. Therefore, the MFFCESSC-SSD can reach a successful balance between accuracy and efficiency.

The MFFCESSC-SSD is evaluated on VisDrone2019 datasets. The result demonstrates the effectiveness of the proposed approach. This method is further extended to practical detection tasks, such as identifying rodents in grassland based on UAV images, demonstrating the superiority of the proposed approach in dealing with small objects in vision tasks.

The contributions of the current study are listed as follows:

(1) A multi-scale feature fusion scheme is designed based on the feature maps of an SSD (i.e., MFF) to add global semantic features to the shallow feature maps, while context mining and self-attentive learning are integrated into an SAU to obtain global contextual information.
(2) Context-enhanced spatial sparse convolution (i.e., CESSC) is performed to reinforce global contextual information and enhance focal features, while a dynamic mask ratio mechanism is proposed that can be automatically updated based on the information of the feature layer.

2. Related Work

2.1. Object Detection in UAV Image

The field of target detection has seen significant advancements; in particular, convolutional neural networks (CNNs) have surpassed traditional object detection methods. CNN-based object detection schemes have been dominated by anchor-based detectors, which can generally be divided into one-stage object detection algorithms based on regression analysis [11–14] and two-stage object detection algorithms based on region suggestion [15–17]. One-stage methods simultaneously complete the classification and localization tasks, making them more efficient than two-stage methods. Meanwhile, two-stage methods refine anchors several times, generating more accurate results than the former, but the training process is more complex and time-consuming [18]. These algorithms utilize the topmost layer of a CNN to recognize objects at multiple scales, which imposes a great burden for a single layer.

Due to the large variations in object size in UAV images, building feature pyramids based on image pyramids is a basic solution. By featuring each level of an image pyramid, online hard example mining (OHEM) produces a multi-scale feature representation in which all levels have rich semantic features, including the high-resolution ones [19]. However, in real applications, this approach is impractical because of its long inference time [20]. The SSD is one of the first attempts employing ConvNets to produce a feature pyramid, which achieves faster detection than other mainstream object detection methods [2]. Recently, numerous studies have optimized feature pyramid networks, allowing them to outperform other methods in the field of target detection based on UAV images [21,22].

In order to mitigate the interference caused by complex backgrounds in UAV images, attention mechanisms are widely used in detection tasks. One notable application of attention mechanisms in UAV target detection is in the development of transformer-based models, such as the foreground enhancement attention Swin transformer network (FEA-Swin), which integrates contextual information into the Swin transformer backbone to improve the accuracy of dense object detection in UAV images [23]. Speeded up robust features (SURFs) is a proposed architecture for feature selection using foveated images that is guided by visual attention tasks and that reduces the processing time required to perform these tasks [24]. The UAV-YOLOv8 model incorporates an attention mechanism called BiFormer to optimize the backbone network, enhancing the model's focus on critical information and improving the detection of small objects in UAV aerial photography scenarios [25]. Similarly, the weather-domain transfer-based attention YOLO model employs an attention mechanism to improve the detection and classification of insulator defects in UAV images, demonstrating the versatility of attention mechanisms in various detection tasks [26].

Overall, the integration of attention mechanisms in UAV target detection models has enhanced detection accuracy and efficiency across various applications and environments.

2.2. Multi-Scale Feature Fusion

Effectively representing and fusing multi-scale features is a major challenge in object detection. Feature pyramid networks (FPNs) build a feature fusion architecture via merging a top–down pathway and lateral connection, combining semantic features and detail information [27]. Scale transformer object networks utilize a scale–transfer layer to obtain shallow features while using a pooling layer to generate feature maps with a large receptive field. These features are directly embedded into the basic detector [28]. The multi-level feature pyramid network (M2Det) [29] leverages a U-shaped module to construct feature pyramids. Efficient object detection (EfficientDet) [30] designs a bi-directional feature pyramid network with cross-scale connection, which reveals that a normalized multi-scale feature fusion approach can achieve a better accuracy and efficiency trade-off.

2.3. Detail of Spatial Sparse Convolution

Spatial sparse convolution involves using a spatial mask to distinguish between 'important' and 'unimportant' regions in the feature maps, assigning a value of 0 to the latter and 1 to the former [31]. The view of UAVs is large, and the foreground containing the target occupies a small part of the entire image. UAV images are inherently a type of spatially non-intensive data, and spatial sparse convolution has been shown to improve the efficiency of processing such data [32]. Given a feature map of size $C \times H \times W$, SSC performs convolution using a shared kernel with dimensions of $C \times 3 \times 3$. The convolution result is a matrix with dimensions of $1 \times H \times W$ which can then be further normalized by a mask matrix. Only those positions with value 1 in the mask matrix will be convolved in the next step. Nevertheless, the mask matrix is derived from a mask ratio $r \in [0, 1]$ according to Gumbel-Softmax [33]. Therefore, the value of the mask ratio determines the efficiency of SSC. Context-enhanced adaptive sparse convolution (CEASC) [34] proposes an adaptive multilayer masking scheme to introduce a mask network for each head, replacing the convolution layer with an SC layer. These methods typically perform inference on one feature map multiple times, limiting their applications on UAV platforms.

3. Method

As shown in Figure 2, ResNet-50 with a residual structure is applied to replace the VGG backbone network in the conventional SSD. ResNet uses skip connections to learn more complex features without degradation and has been shown to improve detection accuracy, especially in challenging scenarios such as detecting small objects or objects with complex backgrounds [35]. MFF aims to fuse feature maps from different layers by applying a lightweight and efficient structure and then generating an FPN. According to the previous study, feature maps with a spatial size smaller than 10 × 10 have limited information for application. Therefore, three different feature layers are selected for concatenation, and the corresponding feature sizes are 38 × 38, 19 × 19, and 10 × 10. CESSC integrates regions of interest with global context through spatial sparse convolution and group normalization. An optimal mask estimation mechanism based on ground-truth labeling is designed to control the activation ratio.

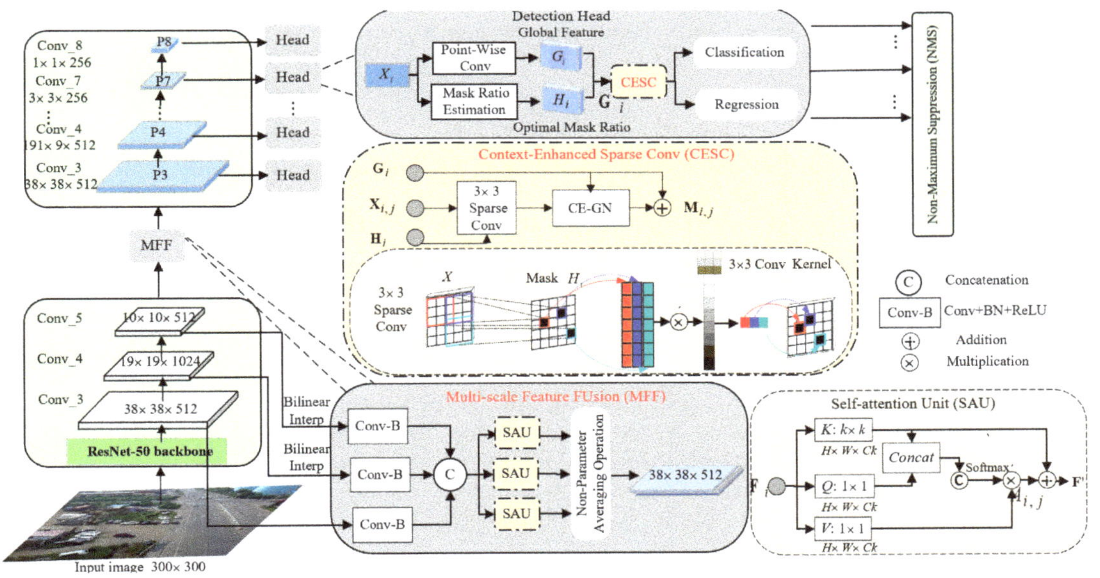

Figure 2. The MFFCESSC-SSD framework based on the SSD (highlighted in green). MFF aims to utilize information from different feature maps and suppress the impact of background noise by using AU blocks; CESSC replaces the detection head in each FPN layer by using a mask feature H_i and a global feature G_i. The mask ratio of H_i is a spatial sparse mask generated from the feature statistics for each layer.

3.1. Multi-Scale Feature Fusion Model

First, bilinear interpolation is used to up-sample the feature maps with sizes smaller than 38 × 38. Conv1 × 1 is then applied to every feature map to reduce the feature dimension. Afterward, batch normalization (BN) and shallow layers (ReLU) are utilized to accelerate the learning speed of the model and alleviate gradient vanishing. The three feature maps are concatenated with the same size in the spatial dimension, and SAU blocks are applied to guide the model to focus on objects in complex backgrounds.

In SAU blocks, the given feature map is $\mathbf{F}_i \in \mathbf{R}^{H\times W\times C_K}$ from the i-th layer of the FPN, where C_K, H, and W refer to the channel size, height, and width of the source feature maps, respectively. $\mathbf{F}_i$ is transformed into queries $\mathbf{Q} = \mathbf{F}_i\mathbf{W}_q$, keys $\mathbf{K} = \mathbf{F}_i\mathbf{W}_k$, and values $\mathbf{V} = \mathbf{F}_i\mathbf{W}_v$ by embedding the matrices ($\mathbf{W}_q, \mathbf{W}_k, \mathbf{W}_v$). Each embedding matrix is generated by conv1 × 1. Notably, the key is encoded using a k × k kernel so that the local contextual information between neighbors can be obtained by Equation (1).

$$\mathbf{R} = \mathbf{K} * \mathbf{Q} \tag{1}$$

The global weighted feature mapping matrix is computed by applying convolutions and the Softmax function to local contextual information, and the formula is expressed as Equation (2).

$$\mathbf{A}_{i,j} = \frac{e^{\mathbf{R}_{i,j}}}{\sum_{j=1}^{H\times W} e^{\mathbf{R}_{i,j}}}, i,j = 1,2,3\ldots H\times W \tag{2}$$

where $\mathbf{A}_{i,j}$ represents the elements of the i-th row and j-th column, all of which make up the global weighted feature matrix $\mathbf{A}$.

$\mathbf{V}$ is subjected to matrix multiplication with matrix $\mathbf{A}$ to obtain a weighted summation across all positions within the same feature map. Subsequently, the resultant matrix multiplication is added to the original series of features to produce the output $\mathbf{F}'$ of the SAUs, as described in Equation (3).

$$\mathbf{F}' = \mathbf{F} + \mathbf{A}\mathbf{V}^T \tag{3}$$

The features $\mathbf{F}'$ extracted from all SAUs undergo a feature fusion process (as shown in Equation (4)). To address the issue of exponential growth in magnitude, a non-parametric averaging operation is employed, which facilitates the derivation of the final fused feature layer.

$$\mathbf{X}_i = \frac{f_i^{1\times1}\mathbf{F}_i' + f_{i+1}^{1\times1}\mathbf{F}_{i+1}' + f_{i+2}^{1\times1}\mathbf{F}_{i+2}'}{3}, i \in n-1 \tag{4}$$

where the i-th layer is comparatively shallower than the i + 1-th layer and the i + 2-th layer, and the $f_i^{1\times1}$ is employed to decrease the number of channels.

The key is encoded using a k × k kernel to address the issue of limited receptive fields in convolutional operations, effectively capturing local contextual information. Concurrently, the self-attention mechanism within each SAU retains the capacity to acquire global contextual information, a characteristic of traditional self-attention, thereby enhancing visual representation. Meanwhile, this study integrates the output of the SAUs into the sparse convolution operation ($\mathbf{F}' = \mathbf{G}_i$) to mitigate the lack of contextual information associated with convolving only the foreground of a UAV image.

3.2. Context-Enhanced Spatial Sparse Convolution

As displayed in Figure 1, global contextual information $\mathbf{G}_i$ from the SAUs is introduced to compensate for the degradation of recognition accuracy due to sparse convolution. Then, CESSC is performed, which takes the feature map $\mathbf{X}_{i,j}$, the mask matrix $\mathbf{H}_i$, and the global feature $\mathbf{G}_i$ as input, where j indicates the j-th SSC layer. $\mathbf{H}_i \in \{0,1\}^{1\times H\times W}$ is generated and formulated according to Equation (5):

$$\mathbf{H}_i \begin{cases} Sigmoid(\frac{\mathbf{S}_i+\varepsilon_1-\varepsilon_2}{\gamma}) > 0.5, & \text{For training} \\ \mathbf{S}_i > 0, & \text{For inference} \end{cases} \tag{5}$$

where $\mathbf{S}_i \in R^{1\times H\times W}$ is a soft feature generated by convolving $\mathbf{X}_i$, and $\varepsilon_1, \varepsilon_2 \in R^{1\times H\times W}$ refer to two Gaussian random noises. As shown in Equation (5), only the region with a mask value of 1 is involved in convolutions, thereby lowering the total computational expense. During inference, the sparsity of Hi is determined by a mask ratio $\gamma \in [0,1]$, which has often been manually set to be higher than 0.9 in recent studies. Considering the different flight altitudes and application scenarios of UAVs, the percentage of meaningful foregrounds in the image varies greatly, and the fixed mask ratio leads to an increase in floating-point operations, while dynamic mask ratios can optimize computational resources. Therefore, Equation (6) is employed to estimate the optimal mask ratio.

$$\gamma = \frac{Fpx(C_i)}{Allpx(C_i)} \tag{6}$$

where $C \in R^{h_i \times w_i \times c}$ is the ground-truth classification result; $Fpx(C_i)$ and $Allpx(C_i)$ indicate the number of pixels belonging to the foreground and that of all pixels, respectively.

The mean value and standard deviation of $\mathbf{G}_i$ are introduced to normalize the feature map after applying SSC to $\mathbf{X}_{i,j}$. The context-enhanced feature $\mathbf{M}_{i,j}$ is defined by context-enhanced group normalization (CE-GN), as expressed in Equation (7).

$$\mathbf{M}_{i,j} = k \times \frac{\mathbf{L}_{i,j} - mean[\mathbf{G}_i]}{std[\mathbf{G}_i]} + b \tag{7}$$

where $\mathbf{L}_{i,j}$ denotes the output after applying SSC; $mean[.]$ and $std[.]$ denote the mean and standard deviation, respectively; and k and b are learnable parameters. By incorporating the global feature $\mathbf{G}_i$, CESSC can better understand the spatial relationships and interactions between objects, leading to improved performance.

In response to the high proportion of small objects in UAV images, $(1-p)^\gamma$ is further added to the cross-entropy, and focal loss (Equation (8)) is defined to guide the algorithm to focus more on positive targets than on backgrounds.

$$FL(p) = -\alpha(1-p)^\gamma \log(p) \tag{8}$$

where p denotes the predicted probability, $\gamma \in [0,5]$ is the focusing parameter used to adjust the rate of change in the weighting factors $(1-p)^\gamma$, and $\alpha \in [0,1]$ is a hyperparameter balancing the contribution of positive and negative samples to the loss.

4. Experiments

First, extensive ablation studies on the object detection task using the VisDrone2019 dataset are conducted to compare the performance of the MFFCESSC-SSD with that of the conventional SSD in terms of accuracy and efficiency and to test the effect of each component. In addition, using this dataset, the MFFCESSC-SSD is compared against other algorithms such as Cascade R-CNN, CenterNet, SyNet, and so on to evaluate the improvement in object detection. Experiments are conducted on the Aerial Rat-Hole dataset (ARH2000), which was acquired by M200 UAVs (DJI-Innovations, Shenzhen, China), for April 2020 and April 2021 to further validate the MFFCESSC-SSD framework.

4.1. Datasets and Evaluation Indicators

VisDrone2019 is a conventional dataset widely used to evaluate algorithms of multi-category object detection based on UAV images. The dataset contains 6471 training images and 548 testing images with a resolution of 1360 × 765 from 10 classes. In the VisDrone2019 dataset, the scales of objects are diverse, and the size of most objects is less than 32 × 32.

ARH2000 is established in this study for monitoring mouse pests. We collected aerial images of five sample plots with different densities of effective gerbil holes in April 2020 and April 2021. We used quadrotor M200 UAVs (DJI-Innovations, Shenzhen, China) as the data acquisition platform and selected clear and cloudless days from 12:00 to 14:00 to minimize the impact of shadows on the classification results. The altitude was set to 30 m, and the forward and side overlaps were set to 70% and 80%, respectively. A total of 445 images were collected, with each sample plot having 89 images of size 5280 pixels × 3956 pixels. We used Pix4Dmapper 4.5.6, a professional drone mapping software, to correct and mosaic the images and generate digital orthophoto maps (DOMs) with a resolution of 0.4 cm. The mouse holes (black patches with specific textures and shapes) in the UAV images are carefully labeled with a labeling tool and then cropped to 300 × 300 to obtain the dataset. This dataset comprises 1728 images for training and 432 images for testing. A considerable amount of black patches in the data, which were formed by withered grass shadows, increases the difficulty of identifying mouse holes.

Average precision (AP) and mean average precision (mAP) are used as evaluation indicators of accuracy, and FPS (frames per second) and model parameters are indicators of efficiency.

4.2. Implementation Details

The network is implemented based on PyTorch 1.7 The network learning rate is set to 0.001, momentum is set to 0.9, and the learning rate decay coefficient is set to 0.1. For VisDrone2019, one image is split into 300 × 300 and independently processed. All models are pretrained on ImageNet to increase efficiency, and the BN layers in the backbone network are frozen during training. In addition, focal loss is introduced to guide the models to focus on small objects. The MFFCESSC-SSD network and other baseline models are trained on NVIDIA RTX 3060Ti GPU.

4.3. Ablation Study

First, every component of MFFCESSC-SSD is evaluated by adopting the SSD as the baseline in all ablation experiments. Table 1 shows that the mAP increases by 3.3% and 2.8% on VisDrone2019 and ARH2000, respectively, by applying the ResNet as a backbone. However, incorporating the enormous parameters of ResNet, the inference speed significantly decreases by 32.7% and 25.3%. The proposed MFF component adds multi-scale semantic features to shallow feature maps and mitigates the impact of the background, which effectively improves the detection accuracy of the model for small objects. Therefore, combining multi-scale feature fusion with an attention mechanism increases the mAP by 13.5% and 5.5% on VisDrone2019 and ARH2000, respectively, while further decreasing the inference speed. The CESSC component reduces the parameters by 39.2% on VisDrone2019 and 52.5% on ARH2000 by adopting global contextual information and SC. Due to the sparsity of targets in the ARH2000 dataset, CESSC more significantly reduces the number of parameters and speeds up inference The CE-GN layer can generate contextual features and global correlations. Thus, the CESSC component further boosts the mAP by 8.7% and 11.2%. Focal loss increases the importance of small objects in the loss function of the MFFCESSC-SSD, and the mAP is slightly boosted by 6.4% and 1.0% on VisDrone2019 and ARH2000, respectively. The introduction of focal loss leads to a small decrease in the model's inference speed due to the fact that the loss function guides the optimization process of the model, and more complex loss functions tend to involve additional parameters or computations.

Table 1. Ablation studies on VisDrone2019 and ARH2000.

Dataset	Baseline	ResNet	MFF	CESSC	Focal Loss	mAP (%)	Parameters/MB	FPS/s
VisDrone2019	✓					19.9	22.9	55
	✓	✓				25.2	39.8	37
	✓	✓	✓			28.6	46.4	31
	✓	✓	✓	✓		31.1	28.2	57
	✓	✓	✓	✓	✓	**33.1**	**28.4**	**56**
ARH2000	✓					78.9	22.9	138
	✓	✓				81.7	39.8	103
	✓	✓	✓			84.2	46.4	92
	✓	✓	✓	✓		93.6	22.2	176
	✓	✓	✓	✓	✓	**94.3**	**22.3**	**176**

4.3.1. Effect of MFF

Figure 3 illustrates the thermal comparison map depicting the UAV image both before and after MFF processing. Initially, the feature information of targets is obscured by significant noise and background data. However, after applying MFF, the feature information is extracted with a much higher accuracy, which indicates that MFF successfully reduced noise and background interference.

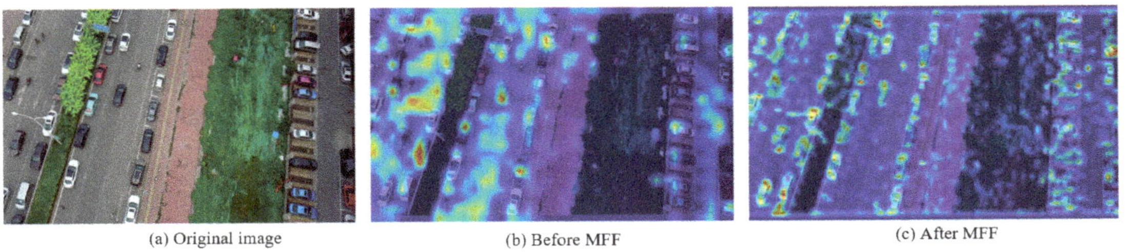

Figure 3. Visualized thermal comparison of UAV images before and after MFF processing.

VisDrone2019 contains numerous small targets. These targets occupy only a few pixels and are characterized by similarity between categories, which leads to poor detection of small targets. The MFF component fuses deep features into shallow feature maps, which increases the accuracy of the model by 5.4%. ARH2000 contains only one type of target (rat hole). The shallow features of the targets, such as shape, color, and texture, are highly prominent. The improved detector achieves an accuracy of 86.2% and runs 3.8 times faster than on the former dataset. The addition of MFF increases the accuracy by 4.5%, which is less than on VisDrone2019. Figure 4 provides a visualization of the two UAV datasets.

4.3.2. Effect of CESSC

SSC only processes convolution in the foreground covered by the mask ratio $\mathbf{H}_i$, which sharply decreases model complexity. Compared to ARH2000, VisDrone2019 has more targets. Therefore, when CESSC is used in the detection head, the inference accelerates by 91.3% and 83.8% on the former and latter datasets, respectively. A precise mask ratio, combined with CE-GN in enhancing the global context, compensates for the loss induced by extracting only foreground features due to SSC. Therefore, the CESSC component further promotes detection accuracy with the two datasets. The above experimental results demonstrate the capability of CESSC to balance accuracy and efficiency.

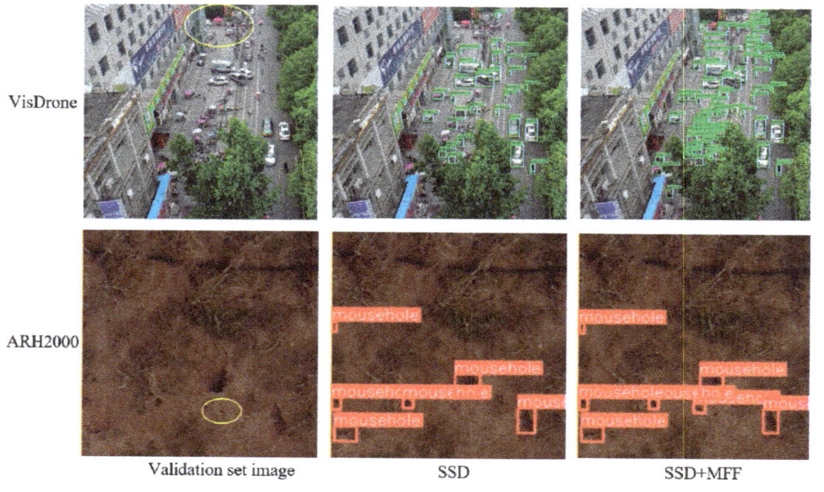

Figure 4. Visualization of detection results. Yellow ovals highlight small objects in validation set, which cannot be detected by SSD. By adding MFF to SSD, model generated denser detection boxes on VisDrone2019 and successfully detected small objects in two datasets.

4.4. Contrast Experiments

The MFFCESSC-SSD is compared with other conventional detectors on VisDrone2019 to further verify the performance of the proposed network, and the results are shown in Table 2. Earlier studies based on VisDrone2019 focused on improving accuracy, including Cascade region-based convolutional neural networks (Cascade R-CNN) [36], CenterNet [36], Synergistic Network (SyNet) [37], and Transformer Prediction Head YOLOv5 (TPH-YOLOv5) [38]. Among them, SyNet is a synergistic architecture that combines Cascade R-CNN and CenterNet using a weighted box. Therefore, SyNet obtains superior results in terms of mAP. YOLO with CSPDarknet as its backbone is widely used and proven to be effective for UAV image object detection. The methods' accuracy increases by 31% and they achieve 16 FPSs with an image size of 416 × 416, which is still below the threshold of 30 FPSs for real-time detection. Notably, model performance is improved with the increase in input size. CESAC and the Self-Attention Guidance and Multi-Scale Feature Fusion-Based Network (SGMFNet) obtain a detection accuracy of 31.7% with high-resolution images. However, the inference speed is still insufficient to meet the requirements of real-time processing. The visualization of the detection results of each algorithm on the VisDrone2019 dataset are shown in Figure 5.

Table 2. Comparison results of different algorithms using VisDrone2019.

Method	Base Detector	Image Size	Backbone	mAP (%)	FPS/s
Cascade R-CNN [36]	R-CNN	960 × 540	ResNet-50	24.7	-
CenterNet [36]	CornerNet	512 × 512	Hourglass-104	14.3	-
SyNet [37]	R-CNN	960 × 540	ResNet-50	26.2	-
TPH-YOLOv5 [38]	YOLOv5	1536 × 1152	CSPDarknet53	31.0	-
Improved YOLOv4 [39]	YOLOv4	416 × 416	CSPDarknet	27.0	16
CEASC [34]	GFL V1	1333 × 800	ResNet18	28.7	22
SGMFNet [40]	-	1536 × 1536	CSPDarkNet53	**31.7**	23
SSD	-	960 × 540	ResNet-50	19.9	55
MFFCESSC-SSD (ours)	SSD	960 × 540	ResNet-50	**33.1**	**56**

Cascade R-CNN, CenterNet, SyNet, TPH-YOLOv5, Improved YOLOv4, CEASC, and SGMFNet are all influential algorithms in the field of UAV image recognition proposed in recent years.

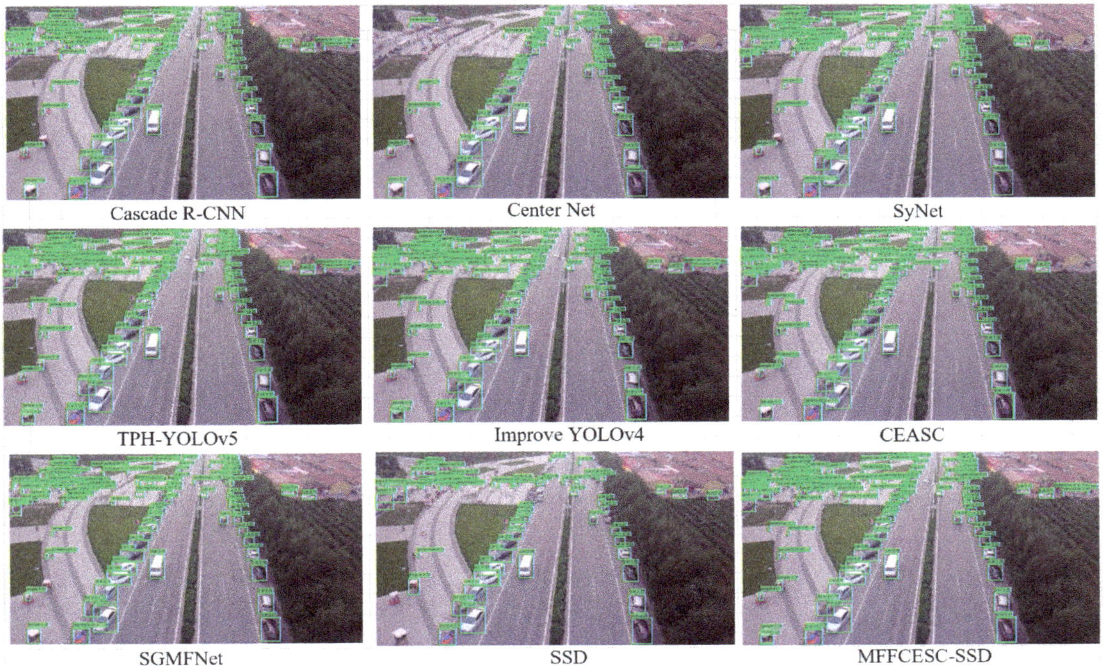

Figure 5. Comparison of detection results of different algorithms using VisDrone2019 dataset. The green boxes are the detected targets, and the MFFCESSC-SSD has the lowest leakage rate.

The SSD with ResNet as its backbone is faster than the above object detection algorithms, but its mAP is only 19.9%. The MFFCESSC-SSD outperforms the conventional SSD by 13.2% in terms of mAP while having limited parameters. The MFFCESSC-SSD is 1.4% more accurate than SGMFNet and runs 1.41 times faster. The contrast experiments reveal that the MFFCESSC-SSD is a competitive algorithm in terms of accuracy and inference speed.

5. Conclusions

A novel and efficient model, namely the MFFCESSC-SSD, is designed in this study to perform object detection in UAV images. A multi-scale feature fusion (MFF) scheme is initially applied to add semantic features to the shallow feature maps and suppress the impact of background noise using self-attention unit blocks. Notably, context mining is integrated into the SAUs to obtain global contextual information. The experiments showed that MFF successfully reduces noise and background interference. CESSC with CE-GN and SSC, which enhances contextual relationships while reducing module complexity, is designed. A dynamic mask ratio mechanism is proposed that can be automatically updated based on the information of the feature layer. Furthermore, CESSC introduces focal loss to address scale changes. As demonstrated in our experiments, our proposed approach yields the highest mAPC and FPS with both VisDrone and ARH2000 when compared to the most recently proposed state-of-the-art object detection algorithms. In conclusion, our approach opens up a new dimension for object detection, i.e., efficient target detection in 2D images using sparse convolution. We hope that this dimension will provide insights for object detection in UAV images for other applications.

Author Contributions: Methodology, Z.Y.; Data curation, J.S.; Writing—original draft, G.Q. All authors have read and agreed to the published version of the manuscript.

Funding: This work has been funded by the National Natural Science Foundation of China (NSFC) under grant No. 52265035, Inner Mongolia Autonomous Region Science and Technology Program under grant No. 2021GG0218, and the Natural Science Foundation of Inner Mongolia Autonomous Region under grant No. 2024LHMS03036.

Institutional Review Board Statement: Not applicable.

Informed Consent Statement: Not applicable.

Data Availability Statement: The original contributions presented in this study are included in the article. Further inquiries can be directed to the corresponding author.

Acknowledgments: The authors would like to thank the anonymous reviewers for their helpful remarks.

Conflicts of Interest: The authors declare no conflict of interest.

References

1. Lin, T.Y.; Maire, M.; Belongie, S.; Hays, J.; Perona, P.; Ramanan, D.; Dollár, P.; Zitnick, C.L. Microsoft coco: Common objects in context. In Proceedings of the Computer Vision–ECCV 2014: 13th European Conference, Zurich, Switzerland, 6–12 September 2014; Proceedings, Part V 13; Springer International Publishing: Cham, Switzerland, 2014.
2. Liu, W.; Anguelov, D.; Erhan, D.; Szegedy, C.; Reed, S.; Fu, C.-Y.; Berg, A.C. Ssd: Single shot multibox detector. In Proceedings of the Computer Vision–ECCV 2016: 14th European Conference, Amsterdam, The Netherlands, 11–14 October 2016; Proceedings, Part I 14; Springer International Publishing: Cham, Switzerland, 2016; pp. 21–37.
3. Zhang, X.; Zhang, Y.; Gao, T.; Fang, Y.; Chen, T. A Novel SSD-Based Detection Algorithm Suitable for Small Object. *IEICE Trans. Inf. Syst.* **2023**, *E106.D*, 625–634. [CrossRef]
4. Li, Z.; Yang, L.; Zhou, F. Fssd: Feature fusion single shot multibox detector. *arXiv* **2018**, arXiv:1712.00960.
5. Lu, X.; Ji, J.; Xing, Z.; Miao, Q. Attention and Feature Fusion SSD for Remote Sensing Object Detection. *IEEE Trans. Instrum. Meas.* **2021**, *70*, 5501309. [CrossRef]
6. Cheng, L.; Ji, Y.; Li, C.; Liu, X.; Fang, G. Improved ssd network for fast concealed object detection and recognition in passive terahertz security images. *Sci. Rep.* **2022**, *12*, 12082. [CrossRef]
7. Graham, B.; Van der Maaten, L. Submanifold Sparse Convolutional Networks. *arXiv* **2017**, arXiv:1706.01307.
8. Yan, Y.; Mao, Y.; Li, B. SECOND: Sparsely embedded convolutional detection. *Sensors* **2018**, *18*, 3337. [CrossRef] [PubMed]
9. Yang, C.; Huang, Z.; Wang, N. Querydet: Cascaded sparse query accelerating high-resolution small object detection. In Proceedings of the 2022 IEEE/CVF Conference on Computer Vision and Pattern Recognition (CVPR), New Orleans, LA, USA, 18–24 June 2022.
10. Song, L.; Li, Y.; Jiang, Z.; Li, Z.; Sun, H.; Sun, J.; Zheng, N. Fine-Grained Dynamic Head for Object Detection. *Adv. Neural Inf. Process. Syst.* **2020**, *33*, 11131–11141.
11. Huang, L.; Yang, Y.; Deng, Y.; Yu, Y. DenseBox: Unifying landmark localization with end to end object detection. *arXiv* **2015**, arXiv:1509.04874.
12. Redmon, J.; Divvala, S.; Girshick, R.; Fahadi, A. You Only Look Once: Unified, Real-Time Object Detection. In Proceedings of the Computer Vision & Pattern Recognition, Las Vegas, NV, USA, 27–30 June 2016.
13. Redmon, J.; Farhadi, A. Yolov3: An incremental improvement. *arXiv* **2018**, arXiv:1804.02767.
14. Zhang, S.; Chi, C.; Yao, Y.; Lei, Z.; Li, S.Z. Bridging the gap between anchor-based and anchor-free detection via adaptive training sample selection. In Proceedings of the IEEE/CVF Conference on Computer Vision and Pattern Recognition, Seoul, Republic of Korea, 16–18 June 2020.
15. Ren, S.; He, K.; Girshick, R.; Sun, J. Faster R-CNN: Towards Real-Time Object Detection with Region Proposal Networks. *IEEE Trans. Pattern Anal. Mach. Intell.* **2017**, *39*, 1137–1149. [CrossRef] [PubMed]
16. Cai, Z.; Vasconcelos, N. Cascade R-CNN: Delving into high quality object detection. In Proceedings of the 2018 IEEE/CVF Conference on Computer Vision and Pattern Recognition, Salt Lake City, UT, USA, 18–23 June 2018.
17. He, K.; Gkioxari, G.; Dollár, P.; Girshick, R. Mask R-CNN. *IEEE Trans. Pattern Anal. Mach. Intell.* **2017**, *42*, 386–397. [CrossRef] [PubMed]
18. Fang, H.; Xia, M.; Zhou, G.; Chang, Y.; Yan, L. Infrared small UAV target detection based on residual image prediction via global and local dilated residual networks. *IEEE Geosci. Remote Sens. Lett.* **2021**, *19*, 7002305. [CrossRef]

19. Shrivastava, A.; Gupta, A.; Girshick, R. Training region-based object detectors with online hard example mining. In Proceedings of the 2016 IEEE Conference on Computer Vision and Pattern Recognition (CVPR), Las Vegas, NV, USA, 26 June–1 July 2016.
20. Girshick, R. Fast r-cnn. In Proceedings of the IEEE International Conference on Computer Vision, Santiago, Chile, 7–13 December 2015; pp. 1440–1448.
21. Zhang, H.; Sun, M.; Li, Q.; Liu, L.; Liu, M.; Ji, Y. An empirical study of multi-scale object detection in high resolution UAV images. *Neurocomputing* **2021**, *421*, 173–182. [CrossRef]
22. Luo, X.; Wu, Y.; Zhao, L. YOLOD: A Target Detection Method for UAV Aerial Imagery. *Remote Sens.* **2022**, *14*, 3240. [CrossRef]
23. Xu, W.; Zhang, C.; Wang, Q. FEA-Swin: Foreground Enhancement Attention Swin Transformer Network for Accurate UAV-Based Dense Object Detection. *Sensors* **2022**, *22*, 14248220. [CrossRef] [PubMed]
24. Gomes, R.B.; Carvalho, B.M.D.; Goncalves, L.M.G. Visual attention guided features selection with foveated images. *Neurocomputing* **2013**, *120*, 34–44. [CrossRef]
25. Wang, G.; Chen, Y.; An, P.; Hong, H.; Hu, J.; Huang, T. UAV-YOLOv8: A Small-Object-Detection Model Based on Improved YOLOv8 for UAV Aerial Photography Scenarios. *Sensors* **2023**, *23*, 14248220. [CrossRef] [PubMed]
26. Liu, Y.; Huang, X.; Liu, D. Weather-Domain Transfer-Based Attention YOLO for Multi-Domain Insulator Defect Detection and Classification in UAV Images. *Entropy* **2024**, *26*, 136. [CrossRef] [PubMed]
27. Lin, T.Y.; Dollár, P.; Girshick, R.; He, K.; Hariharan, B.; Belongie, S. Feature pyramid networks for object detection. In Proceedings of the IEEE Conference on Computer Vision and Pattern Recognition, Honolulu, HI, USA, 21–26 July 2016.
28. Zhou, P.; Ni, B.; Geng, C.; Hu, J.; Xu, Y. Scale-transferrable object detection. In Proceedings of the IEEE Conference on Computer Vision and Pattern Recognition, Salt Lake City, UT, USA, 18–22 June 2018.
29. Zhao, Q.; Sheng, T.; Wang, Y.; Tang, Z.; Chen, Y.; Cai, L.; Ling, H. M2Det: A single-shot object detector based on multi-level feature pyramid network. In Proceedings of the AAAI Conference on Artificial Intelligence, Honolulu, HI, USA, 27 January–1 February 2019.
30. Tan, M.; Pang, R.; Le, Q.V. EfficientDet: Scalable and efficient object detection. In Proceedings of the IEEE/CVF Conference on Computer Vision and Pattern Recognition, Seattle, WA, USA, 13–19 June 2020.
31. Wang, L.; Dong, X.; Wang, Y.; Ying, X.; Lin, Z.; An, W.; Guo, Y. Exploring Sparsity in Image Super-Resolution for Efficient Inference. In Proceedings of the 2020 IEEE/CVF Conference on Computer Vision and Pattern Recognition (CVPR), Nashville, TN, USA, 20–25 June 2020.
32. Graham, B.; Engelcke, M.; van der Maaten, L. 3D Semantic Segmentation with Submanifold Sparse Convolutional Networks. In Proceedings of the 2017 IEEE/CVF Conference on Computer Vision and Pattern Recognition, Salt Lake City, UT, USA, 18–22 June 2017.
33. Verelst, T.; Tuytelaars, T. Dynamic Convolutions: Exploiting spatial sparsity for faster inference. In Proceedings of the 2020 IEEE/CVF Conference on Computer Vision and Pattern Recognition (CVPR), Seattle, WA, USA, 14–19 June 2020.
34. Du, B.; Huang, Y.; Chen, J.; Huang, D. Adaptive Sparse Convolutional Networks with Global Context Enhancement for Faster Object Detection on Drone Images. In Proceedings of the 2023 IEEE/CVF Conference on Computer Vision and Pattern Recognition (CVPR), Vancouver, BC, Canada, 18–22 June 2023.
35. Fu, C.Y.; Liu, W.; Ranga, A.; Tyagi, A.; Berg, A.C. DSSD: Deconvolutional Single Shot Detector. *arXiv* **2017**, arXiv:1701.06659.
36. Cao, Y.; He, Z.; Wang, L.; Wang, W.; Yuan, Y.; Zhang, D.; Zhang, J.; Zhu, P.; Van Gool, L.; Han, J.; et al. VisDrone-DET2021: The vision meets drone object detection challenge results. In Proceedings of the IEEE/CVF International Conference on Computer Vision, Montreal, BC, Canada, 11–17 October 2021; pp. 2847–2854.
37. Albaba, B.M.; Ozer, S. SyNet: An Ensemble Network for Object Detection in UAV Images. In Proceedings of the 2020 25th International Conference on Pattern Recognition (ICPR), Milan, Italy, 10–15 January 2021.
38. Zhu, X.; Lyu, S.; Wang, X.; Zhao, Q. TPH-YOLOv5: Improved YOLOv5 Based on Transformer Prediction Head for Object Detection on Drone-captured Scenarios. In Proceedings of the IEEE/CVF International Conference on Computer Vision (ICCV) Workshops, Montreal, BC, Canada, 11–17 October 2021.
39. Ali, S.; Siddique, A.; Ateş, H.F.; Güntürk, B.K. Improved YOLOv4 for aerial object detection. In Proceedings of the 29th Signal Processing and Communications Applications Conference (SIU), Istanbul, Turkey, 9–11 June 2021; pp. 1–4.
40. Zhang, Y.; Wu, C.; Zhang, T.; Liu, Y.; Zheng, Y. Self-Attention Guidance and multi-scale Feature Fusion-Based UAV Image Object Detection. *IEEE Geosci. Remote Sens. Lett.* **2023**, *20*, 6004305.

applied sciences

MDPI

Article

Increasing the Classification Achievement of Steel Surface Defects by Applying a Specific Deep Strategy and a New Image Processing Approach

Fatih Demir [1,*] and Koray Sener Parlak [2]

[1] Software Department, Engineering Faculty, Firat University, Elazig 23119, Turkey
[2] Electric and Electronic Department, Firat University, Elazig 23119, Turkey; kparlak@firat.edu.tr
* Correspondence: fatihdemir@firat.edu.tr

Abstract: Defect detection is still challenging to apply in reality because the goal of the entire classification assignment is to identify the exact type and location of every problem in an image. Since defect detection is a task that includes location and categorization, it is difficult to take both accuracy factors into account when designing related solutions. Flaw detection deployment requires a unique detection dataset that is accurately annotated. Producing steel free of flaws is crucial, particularly in large production systems. Thus, in this study, we proposed a novel deep learning-based flaw detection system with an industrial focus on automated steel surface defect identification. To create processed images from raw steel surface images, a novel method was applied. A new deep learning model called the Parallel Attention–Residual CNN (PARC) model was constructed to extract deep features concurrently by training residual structures and attention. The Iterative Neighborhood Component Analysis (INCA) technique was chosen for distinguishing features to lower the computational cost. The classification assessed the SVM method using a convincing dataset (Severstal: Steel Defect Detection). The accuracy in both the binary and multi-class classification tests was above 90%. Moreover, using the same dataset, the suggested model was contrasted with pre-existing models.

Keywords: steel surface defects; PARC model; classification

1. Introduction

Improving productivity and preserving product quality are crucial in production systems [1]. One of the most crucial aspects of quality control is the detection of surface defects. Traditional detection techniques rely on expert inspection for control, which has disadvantages such as time wastage, poor productivity, and poor reliability [2]. Specifically, automatic surface defect detection is becoming increasingly important in steel industries having extensive production fields. Due to heat treatment, non-metallic inclusion, corrosion, and emulsification, flat-rolled steel can develop flaws such as cracks, pitted surfaces, scratches, and patches [3]. Evaluating defect geometry and generating a sizable sample of statistical data is crucial for improving the surface defect process. As a consequence, automated detection and classification systems can prevent unforeseen equipment failure.

Automatic surface defect detection has frequently been accomplished using traditional machine learning and deep learning-based techniques [4]. Conventional classification approaches including model-based, statistical-based, and spectral-based methods typically rely on statistical data and visual features. These techniques, such as thresholding, Sobel, Local Binary Patterns (LBPs), Fourier transform, and Canny's algorithm, are used to convert

https://doi.org/10.3390/app15084255

features obtained manually, but factors including the background, lighting, and camera angle directly impact the effectiveness of defect identification [5]. Additionally, these techniques have limitations when used on various surfaces, making them unsuitable for use in practical situations [6]. Deep learning-based methods are now used to improve defect-detection capabilities in computer vision [7]. When using deep learning techniques, the algorithm can function by producing prompt and precise predictions even in the absence of supervision. Thus, labor and time saving by detecting automatically in the steel factories have made important advantages in comparison with manual control by technicians and engineers [8].

2. Related Studies

The number of studies using machine learning to identify surface defects of steel has increased during the past few years. Martins et al. [9] performed automatic surface fault detection of rolled steel using artificial neural networks with image processing. The classification of certain defects, including clamps, holes, and welding, was detected by the image analysis Hough Transform method, while other complicated defects, such as exfoliation, oxidation, and waveform, were detected by applying Self-Organizing Maps and Principal Component Analysis to extract features. The system achieved an overall accuracy of 87% after managing real-world datasets. Li et al. [10] designed a feature fusion-based method to improve steel surface defect detection in their proposed model. In this method, a multiscale feature extraction (MSFE) strategy is adopted from a YOLO-based model. With the MSFE algorithm, features with different scales are extracted from multidimensional kernels of different convolution layers. An efficient feature fusion technique is then used to maximize feature discriminability. This model, developed on the publicly available NEU-DET dataset, achieved an optimum accuracy of 73.08%. Pang and Tan [11] developed a graph neural network-based method for detecting steel surface defects. They also used a novel attention mechanism called HDmA in this approach. This strategy was also successful in detecting defects in different fields of view. This method was tested on NEU-DET and GC-10 datasets and achieved 79.04% and 66.93% accuracy, respectively. Zhang et al. [12] presented a model based on YOLO v5 for detecting defects on steel surfaces. A multi-feature fusion technique, Res2Attention blocks, was used to improve the performance of the model. Model performance was tested on the NEU-DET and GC10-DET datasets. The classification accuracies were 78.5% and 67.3%, respectively. A vision-based automatic detection method with three-section defect detection, region extraction, and industrial liquid quantification was proposed by Zhao et al. [13]. They discovered that industrial liquids were measured with an accuracy rate of 90%. The accuracy rate of the recognition of cracks and scratches was obtained as 91% or more. By creating a dataset with six defects: scars, scratches, inclusions, burrs, seams, and iron scales. Li et al. [14] explored the surface defect recognition of steel strips by converting You Only Look Once (YOLO) completely into convolutional layers. The rates of detection, recall, and mAP were 99%, 95.86%, and 97.55%, respectively. Fu et al. [15] applied the SqueezeNet model by pre-training on the dataset of ImageNet for the classification of six different defects, which are rolled-in scales, inclusions, patches, scratches, crazing, and pitted surfaces. They came to the conclusion that the learned features significantly outperformed the hand-crafted ones for datasets that were heterogeneous and unseen. By using the strong time-sequenced properties of defects, Liu et al. [16] studied steel defect detection periodically by using a convolutional neural network (CNN) and a long short-term memory (LSTM) to detect roll mark defects. After extracting the defect features of samples with CNN, vectors were added to the long- and short-term memory to detect defects. As a result, the proposed method reached 81.9% of the performance of defect detection and outperformed the CNN approach. The performance of the system was raised to a rate of 6.2% by enhancing the attention mechanism. Liu et al. [17] utilized the dataset of NEU-CLS to improve the concurrent convolutional neural network (ConCNN) with light weight by using different scales of samples. The method's accuracy performance was found to be 98.89% with a duration of 5.58 ms. The classification pre-training of surface defects was conducted using VGG19 by Guan et al. [18]. Feature extraction from the various levels of the defect weight model was performed using VGG19. Following that, SSIM and a Decision Tree were used to estimate the structure of VGG19 and the quality of the feature picture. The experiment used a dataset from Northeast University with six different types of steel surface defects, crazing, inclusions, patches, pitted surfaces, rolled-in scales, and scratches, each with 300 samples and a total of 1800 grayscale images. Following 23,000 steps, it was discovered that the VSD model's validation rate was 89.86% higher than that of ResNet and VGG19. By using U-net and Deep Residual techniques for the classification of four different defects, Amin and Akhter [19] concluded that their system performances were 0.731 and 0.543 in terms of the Dice coefficient accuracy, respectively. In total, 12,568 training images and 1801 test images were created with a $1600 \times 256 \times 1$ image size. A privacy dataset was explored by Zhao et al. [3] utilizing enhanced Faster R-CNN. By enhancing the conventional Faster R-CNN algorithm, the network structure of the Faster R-CNN was rebuilt. Following the testing, upgraded Faster R-CNN was proven to have a greater mean average accuracy performance for crazing, inclusions, patches, rolled-in scales, and pitted surface defect types with the following values of 0.501, 0.791, 0.792, 0.905, and 0.649, respectively.

After a detailed search of datasets studied at the academic level, it was understood that studies of classification are limited to surface defect detection and classification. However, surface flaws in real-world practices often affect only a small portion of the overall steel surface. As a result, using such data makes success more challenging. Because the Steel Surface Defect Database is a difficult dataset, a deep learning-based strategy that will perform well with it is used in this work.

3. Literature Gap, Motivation, and Contributions

To summarize the literature in general, in terms of deep learning, pre-trained models or lightweight CNN models have been used to classify defects on steel surfaces. Additionally, no frequency–time conversion algorithm has been used to increase the discrimination in the images. Looking at the literature, it is evident that many deep learning-based studies have been conducted to minimize defects in steel production. In studies focusing on the classification problem, the classification accuracy ranged between 80% and 90%. Especially in YOLO-based fault detection studies, the classification accuracy performance was between 65% and 75%. When these accuracy values are considered, it is obvious that the error performance of steel surfaces is still open to improvement. In a steel mill with a high production capacity, even a 2–3% increase in the defect detection performance will significantly improve the production quality. Therefore, there is a serious need for a deep learning-based application that detects steel surface defects with high performance.

This study was carried out to distinguish and classify defects on steel surfaces, as a small improvement in defect detection in high-capacity steel production will provide a large number of defect-free steel products. The proposed approach is designed in four main stages. The first stage is the processing of raw images. The second stage is responsible for extracting features from the Parallel Attention–Residual CNN (PARC) model. The third stage involves feature selection with the Iterative Neighborhood Component Analysis (INCA) algorithm. The final stage includes classification with a powerful algorithm. The proposed approach has three important contributions, as outlined below.

- Attention and residual structures in the CNN model were added to the PARC model and trained in parallel. This increased the representation power of the features extracted from the PARC model. Therefore, the classification performance was improved. Although the parallel integration of attention and residual blocks has been studied in previous works, our proposed PARC architecture introduces a distinct combination specifically tailored for defect recognition in steel surface images. To validate its superiority, we conducted comparative experiments against baseline CNNs with only residual blocks, only attention modules, and sequential Attention–Residual designs. The results demonstrate that PARC achieves a consistently higher accuracy and robustness across both multi-class and binary classification tasks.
- Raw images were processed with a new approach. In this approach, 1D stack data are used instead of a signal to obtain spectrogram images. For 1D stack data, the gradient of the raw images was taken and then converted into sequential 1D stack data. The gradient operation highlighted the differences in pixel values in the image. As a result, the classification performance is improved with this pre-processing procedure. Existing methods all use pixel data. In this study, thanks to the transformation of the image into 1D stack data, it was possible to access the frequency information of images that form a pattern with each other, such as scratches and cracks. In this case, it increased the classification performance compared to raw images.

4. Dataset

The suggested methodology was tested using a challenging dataset from the Kaggle database called Severstal: Steel Defect Detection (2019) [20]. Two tasks involving binary and multi-class classification used the dataset. Images of steel faults (6666 images) and images of no defects (5902 images), obtained using specialized imaging equipment to identify defects, were employed in the binary classification. The multi-class classification work used 6668 steel defect images with pitted surfaces, crazing, scratches, patches, and multi-defect class labels. Each sample of the collection was saved as a 1600 × 256 JPG image. To lessen the hardware requirements for this study, each sample was re-saved in a size of 200 × 32. Figure 1 shows a few samples for each class.

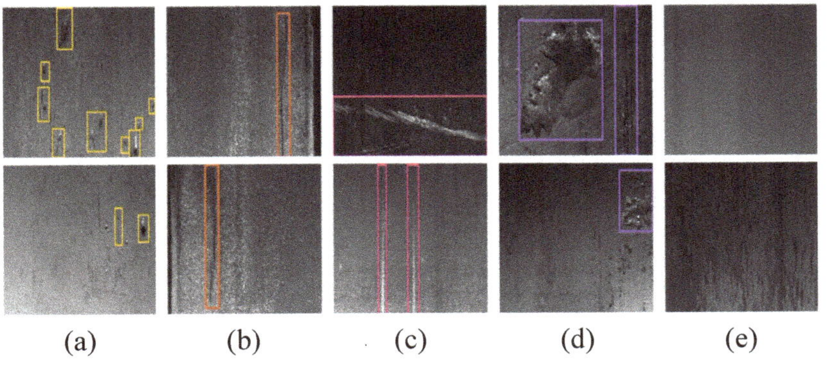

Figure 1. Dataset samples of classified shapes with defects: (**a**) pitted surface, (**b**) crazing, (**c**) scratches, (**d**) patches, (**e**) non-defect [21].

5. Proposed Methodology

The methodology of this study introduces a novel and robust approach to enhance the classification performance for surface defect detection, consisting of five distinct yet interconnected stages. Figure 2 shows the representation of the proposed approach. The innovative aspect of the approach begins with the pre-processing of raw images, where a unique method is employed. Unlike traditional image pre-processing techniques, this study applies a spectrogram algorithm to time series signals derived from images that contain surface defects. This step transforms 2D image data into a 1D stack by utilizing pixel values, enabling the extraction of spectrogram images. This novel transformation provides a powerful representation of pixel variations in defect regions, enhancing the model's ability to capture subtle details crucial for classification.

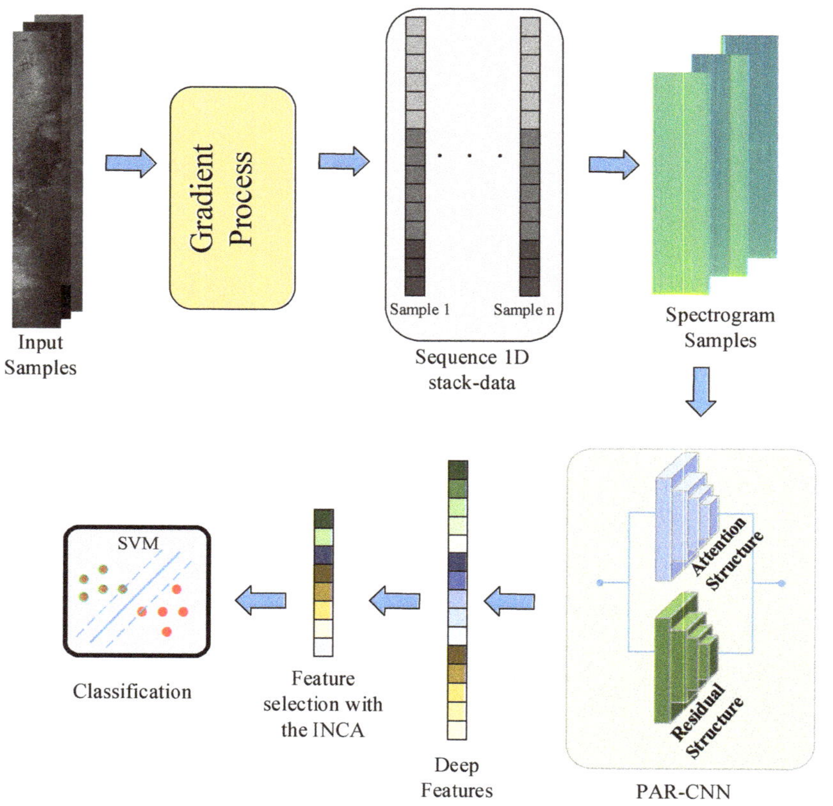

Figure 2. The framework of the proposed methodology.

In the second stage, the processed spectrogram images are used to train a newly designed architecture known as the Parallel Attention–Residual CNN (PARC) model. Figure 3 shows the layer connections of the PARC model. The innovation in this model lies in its dual use of attention and residual mechanisms within a customized CNN framework, operating in parallel. Attention mechanisms focus on emphasizing critical areas within the image, while residual connections allow the preservation of feature information from earlier layers by passing it forward to subsequent layers. This synergy ensures that vital details, which may have otherwise been diminished or lost in deep layers, are retained and utilized, thus improving the learning capability of the network. While state-of-the-art architectures such as Swin Transformer, EfficientNetV2, and YOLOv7 have demonstrated remarkable performance across a variety of vision tasks, they are primarily designed as general-purpose models and often involve high computational costs and large parameter counts. In contrast, the proposed PARC (Parallel Attention and Residual Convolution) model is specifically tailored for steel surface defect recognition, focusing on enhancing subtle and fine-grained texture variations that are commonly present in such industrial inspection tasks. The parallel integration of attention and residual mechanisms in PARC enables it to capture both local defect features and global structural information more effectively, without significantly increasing the computational complexity. Compared to Transformer-based models like Swin Transformer, which require large datasets and extensive training resources to generalize well, PARC achieves a competitive or superior accuracy with a lightweight and task-optimized architecture. Furthermore, unlike YOLOv7, which is object-detection-focused and might exhibit overkill or be less efficient for fine-grained classification tasks, PARC offers a better balance between precision, speed, and model complexity. Its compact design makes it more suitable for real-time deployment in edge environments commonly found in industrial inspection systems. Detailed layer information of the PARC model is given in Table A1 in the Appendix A section.

The third stage focuses on feature extraction from the fully connected (FC) layers of the PARC model, where deep features are drawn from both the activations and the processed input data. This stage sets the foundation for the next key novelty: instead of relying solely on the softmax classifier typically employed in CNN models, the extracted features are evaluated using a range of highly effective classification algorithms. This offers a fresh perspective on model training, allowing for a more versatile comparison of classifiers.

In the fourth stage, an efficient feature selection algorithm called INCA is applied. INCA stands out for its ability to significantly reduce computational costs while simultaneously improving classification performance. This step ensures that only the most relevant and influential features are retained, minimizing the overhead associated with large datasets and complex models. The Iterative Neighborhood Component Analysis (INCA) technique offers distinct advantages over more traditional feature selection and dimensionality reduction methods such as Principal Component Analysis (PCA) and Minimum Redundancy Maximum Relevance (mRMR).

Unlike PCA, which performs unsupervised dimensionality reduction by projecting data onto directions of maximum variance regardless of class labels, INCA is a supervised method that directly optimizes class separation by maximizing the classification accuracy. This makes INCA particularly effective in tasks where discriminative power is more important than variance preservation, such as defect detection or fine-grained classification.

Compared to mRMR, which focuses on selecting features that are most relevant to the target variable while minimizing redundancy between features, INCA iteratively refines the feature subset based on its impact on classification performance. This iterative refinement process allows INCA to dynamically adapt to the dataset structure and learn an optimal feature subset tailored to the classifier used, which often leads to a higher accuracy and better generalization.

In summary, while PCA and mRMR are powerful and widely used, INCA provides a more targeted and performance-driven approach to feature selection by integrating label information and classifier feedback during feature optimization.

Finally, in the fifth and final stage, this study implements seven popular classification algorithms, including Decision Tree (DT), Support Vector Machine (SVM), K-Nearest Neighbor (KNN), Naïve Bayes (NB), Linear Discriminant (LD), Subspace KNN, Subspace Discriminate, and RUSBoosted Trees. Through extensive testing, the SVM algorithm demonstrated the best performance, highlighting the efficiency and accuracy of this method in classifying surface defects. This comprehensive methodology not only demonstrates significant advancements in pre-processing and model design but also introduces innovative techniques in feature selection and classifier evaluation, making it a substantial contribution to the field of surface defect detection and classification.

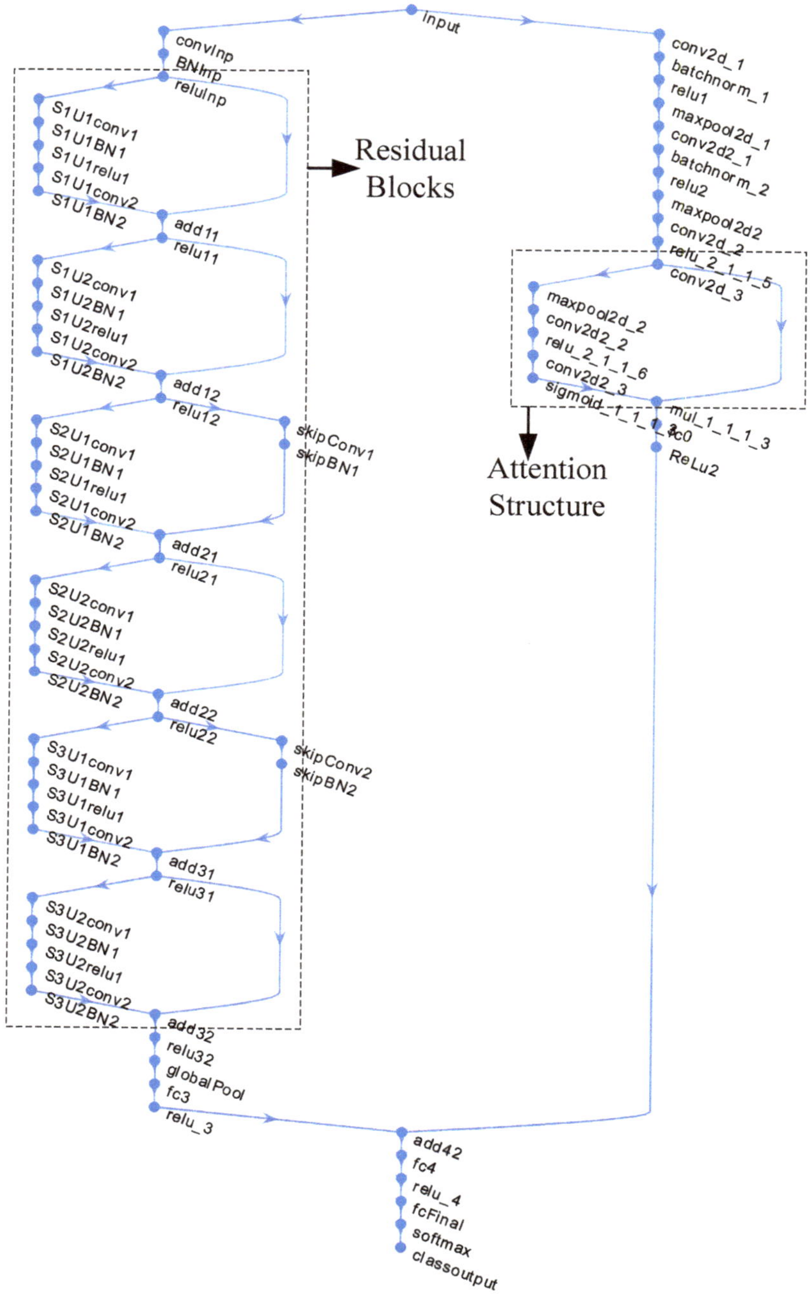

Figure 3. The layer structure of the proposed PARC model.

6. Images-to-Spectrogram Image Transform

Obtaining image-to-image spectrogram images for surface defect detection offers a powerful way to leverage frequency information, enhance feature representation, reduce noise, and enable the application of advanced analysis techniques, ultimately leading to more accurate and reliable defect detection.

The primary motivation for using the spectrogram lies in its ability to represent localized frequency variations over space, analogous to how time–frequency analysis is used in signal processing. Surface defects, although visually subtle, often introduce local structural irregularities that manifest as distinct frequency patterns when observed in a gradient-enhanced 1D representation. By transforming these data into a spectrogram, we can effectively capture these localized textural anomalies in a way that traditional spatial or frequency domain techniques might overlook.

Moreover, spectrograms offer a dual representation—combining gradient magnitude variations with localized frequency content—which enhances the model's ability to distinguish fine-grained defect features. This is particularly beneficial in scenarios where defects are embedded within noisy or highly textured steel surfaces.

One of the most popular techniques in the field of signal processing is the processing of signals in the time–frequency domain. The most crucial information reveals how and when the spectral information of the signals analyzed in the time–frequency domain changes [22]. In procedures that are linearly time-invariant (LTI), like Fourier transforms, such information cannot be seen. The optimum method for observing a signal's spectrum information in the time–frequency domain is through the use of a spectrogram. The spectrogram, which employs a sliding window to determine the Fourier transform of the signal, is often employed in the spectrum analysis of many non-stationary signals, including biological, voice, music, and seismic data [23].

The graph showing the change in 1D stack data with the gradient applied for each class is presented in Figure 4. As observed in this figure, the gradient operation yields different amplitude values depending on the type of surface defect. Particularly in the non-defect class, the gradient output values are minimal, indicating little change across the dataset for this class.

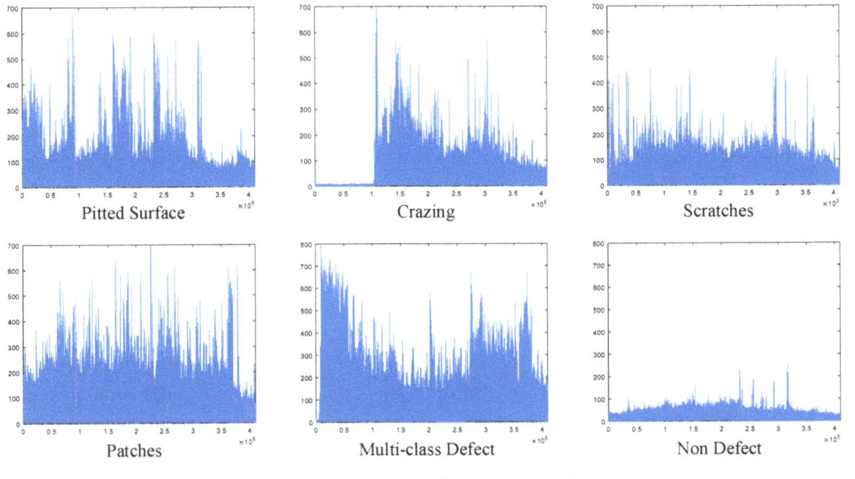

Figure 4. One-dimensional stack data with the gradient operation for each class.

In the final step, spectrogram images were generated from the 1D stack data. The Hamming window was selected for signal windowing due to its advantages, especially its narrow main lobe and rapidly decaying side lobes in the frequency domain. These features help prevent spectral leakage, which can weaken the accuracy of spectral information. A sample size of 512 was chosen for the window, and the overlapping ratio was set at 0.125.

This overlap helps reduce data loss in the 1D stack, but it also increases the computational load required to produce the spectrogram images.

To enhance the clarity of the spectral data, special attention was given to amplitude values while generating the spectrogram images. A threshold was applied to filter out points with low amplitude values, as high-amplitude points are generally more reliable in terms of spectral information. This approach simplifies the images, which is beneficial when using neural network classifiers, as less complex images tend to improve the classification performance. A 512-point FFT length was applied to ensure consistency. The resulting spectrogram images for each class are displayed in Figure 5, where distinct patterns can be observed for each class, demonstrating the effectiveness of this approach for capturing class-specific spectral features. The colormap viridis option was selected for spectrogram coloring.

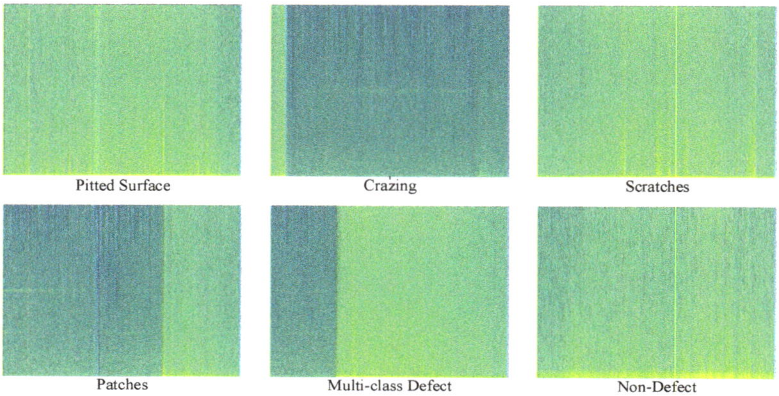

Figure 5. Conversion samples from raw images to spectrogram images for each class.

Using colormap options like viridis in spectrogram visualization offers several advantages. Viridis provides perceptual accuracy, as it is designed with evenly spaced color differences that allow users to distinguish variations in data more easily. It also maintains readability in monochrome formats, ensuring that important information is preserved even when printed or viewed in low-resolution settings. The smooth, linear color transitions of Viridis avoid abrupt shifts, making it easier to interpret differences in intensity or frequency. Additionally, Viridis ensures that small changes in data are accurately represented, reducing the risk of misinterpretation, which is particularly important for scientific visualizations.

As shown in Figure 5, there is sudden color darkening or sudden appearance and disappearance of color lines in all classes except the class with no error. This shows that the Viridis option is ideal for spectrogram transformation.

The convolution layer of the CNN method intends to extract characteristic features processing input samples with convolution filters [24]. The mathematical computing of two functions is described as convolution. By conducting element-wise multiplication on each element, the convolution computing in the CNN technique implements a filter or kernel function utilized for the raw data's transformation into processed data [25,26]. Each window's output in a shift operation is determined by the total multiplication of the element information [27,28].

The batch normalization (BN) technique is used to design a more regular convolutional neural network. Moreover, during training, the CNN extinction gradient becomes more resistant [29,30].

In the Rectified Linear Unit layer (ReLU), the full "f(k) = max(0, k)" formula is used for all inputs as the layer activation function [31,32]. ReLU's derivative is more appropriate and performs faster for algorithms like backpropagation as it is simpler than the sigmoid function.

The softmax function is typically used for the output in deep learning models [33]. The function converts the class scores from the fully connected layer to probabilistic values ranging from 0 and 1. The softmax function is denoted by $S(a_j)$ in Equations (1) and (2). It obtains an N-dimensional input vector, then produces a subsequent input vector with N-dimensional, having values between 0 and 1 [34,35]. Additionally, even though the softmax function is frequently chosen for the output layer in deep learning models, an SVM classifier can be preferred [36]. The exponential feature of the softmax function makes the differences across classes more certain.

$$S(a_j) : \begin{bmatrix} a_1 \\ \cdots \\ a_n \end{bmatrix} \rightarrow \begin{bmatrix} s_1 \\ \cdots \\ s_n \end{bmatrix} \tag{1}$$

$$S(a_j) = \frac{e^{a_j}}{\sum_{k=1}^{n} e^{a_k}} \tag{2}$$

Figure 6 represents the attention module utilized in this research. The gating signal vector, symbolized by g_i , has a wider scale at the feature map of output for ith layer "(x_i)", determining the focus region for each pixel [37]. Equations (3) and (4) give a computed output by applying element-wise multiplication.

$$output = \alpha_i \times x_i \tag{3}$$

$$\alpha_i = \sigma\left(\varphi^T\left({w_x}^T x_i + {w_g}^T g_i + b_g\right) + b_\varphi\right) \tag{4}$$

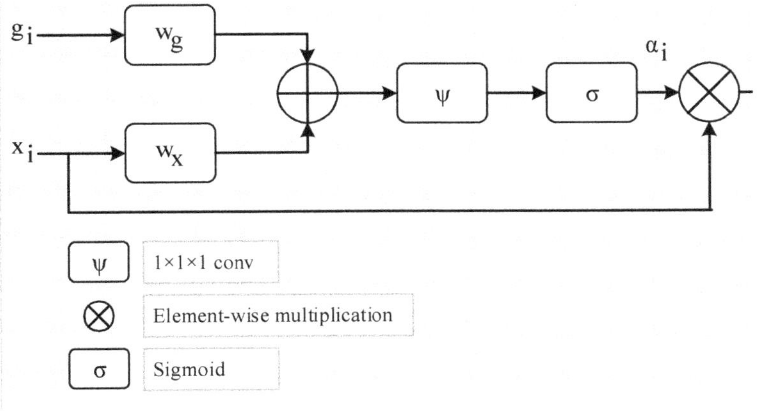

Figure 6. The illustration for the proposed attention layer.

The linear transformations w and φ using the $1 \times 1 \times 1$ dimensional convolution operator are the bias terms b_φ and b_g, respectively. The weights of the attention modules are adjusted randomly at first when the entire deep architecture is trained from end to end [38].

In conventional neural networks, each layer provides information to the next layer. Each layer enhances the subsequent layer directly for the network having residual blocks, then advances to layers, which are two to three hops away [39]. In multi-layer networks, the gradient vanishing problem is decreased by residual blocks. The two significant points of residual blocks are given below.

- The inclusion of new layers would not harm the model performance because regularization will disregard them if not necessary.
- When incoming layers are convenient, layer weights or kernels are not "0" because of present regularization. Therefore, the model performance can slightly improve.

The training of the attention mechanism and the residual blocks were applied concurrently in parallel in the developed PARC model. The main target is to transfer feature maps in residual and attention structures into one feature map. Hence, during the training phase, optimization determines the characteristic values from every two structures obtained from the feature map.

7. INCA Algorithm

Increased system speed without reducing approach success is the most important aim of feature selection techniques. The literature on machine learning algorithms has been enhanced by several feature selection methods [40–42]. Specifically, the feature selection techniques reduce the computational cost of the deep learning algorithms having a lot of features. The performance of the feature selection approach on the feature set should be profoundly assessed [43]. However, in deep learning methods, the application of this analysis method is not time-saving. While the Minimum Redundancy Maximum Relevance (mRMR) technique performs better by using a non-parametric feature set, Principal Component Analysis and the Linear Discriminant analysis techniques outperform by applying a linear feature set [44]. For classification problems, current research mostly depends on feature extraction-based algorithms in terms of feature weight relations [45–47].

The NCA is one of the most popular features of importance-based selection algorithms since they offer a variety of classification strategies. Additionally, the computational time for these approaches outperforms the PCA and mRMR algorithms.

Neighborhood Component Analysis (NCA), a method for feature selection and dimensionality reduction that is widely used in classification studies, is one of the most reliable supervised learning approaches for classifying multidimensional data into distinct classes [48,49]. The classification tasks performed by NCA are carried out with learning vector optimization criteria related to the categorization accuracy performance of the nearest neighbor classifier. In particular, a linear projection chosen by NCA maximizes the projected area's performance of the nearest neighbor classifier. In NCA, training data with related class labels are applied to choose the projection that divides the classes effectively in the detected area. Nevertheless, the NCA makes assumptions about the distribution of each class, which are not reliable. It offers an equivalent fit to Gaussian mixtures for distribution modeling. To maximize the objective function $F(w)$ for w, the regularized objective function is used in Equation (5).

$$F(w) = \frac{1}{n}\sum_{i=1}^{n} P_i - \lambda \quad (5)$$

where the overall sample size is n, the value of the probability of ith the specimen is "P_i", the parameter of regularization is "λ", the dimension of the feature is "p" and the weight of the feature is "w_r". The weight values for the feature may be very near to "0" in case the selection of "λ" is performed at random. The relevant features have no importance for the method when weights are very near to zero. Thus, the parameter λ has to be arranged.

Iterative Neighborhood Component Analysis (INCA) holds advantages over Neighborhood Component Analysis (NCA) primarily due to its adaptive feature selection mechanism. Unlike NCA, INCA iteratively selects features, allowing it to dynamically adjust and optimize the learning process by choosing the most relevant features for a given task. This adaptability enhances the discriminative power, robustness, and generalization performance. By reassessing and modifying feature selections during each iteration, INCA is better equipped to handle diverse datasets and mitigate the impact of noisy or irrelevant features. The flexibility offered by INCA in tailoring its approach to specific dataset characteristics makes it a promising choice for various machine learning applications.

Choosing the parameter λ randomly may not give the best feature selection result. Here, the most reliable way to select the lambda parameter is to use an optimization algorithm. The Stochastic Gradient Descent (SGD) algorithm provides exemplary performance in many optimization problems. In this study, the lambda parameter used in INCA was selected with the SGD optimization algorithm. The pseudocode expression of this algorithm (INCA) is given in Algorithm 1.

Algorithm 1. Pseudocode of the PARC model

Inputs: features from the PARC model, labels,
Output: the selected feature (features_out)

```
features_out = INCA_algorithm (features, labels)
begin
  nca = fscnca (Xtrain, ytrain, 'sgd', bestlambda);
      for i = 1 to N do
          search the best lambda parameter by using the NCA
      end for i
      compute feature weights with the best lambda
      indeces = weights (indices)
          for j = 1 to length (features) do
              if weights (j) >= threshold
                  append j to the new indices list
              end if
          end for j
      features_out = fea[new_indeces]
      return features_out
end
```

The parameter N is the number of optimization iterations and is set to 20 (default value). The threshold value is used to eliminate features including low-weight values. For the binary and multi-class classification, threshold values were selected as 0.5 and 0.2, respectively.

8. Experimental Studies

The computer used in the experimental studies has a 4 GB graphics card, i7 intel 5500U processor (Intel, Santa Clara, CA, USA), and 16 GB RAM, and the MATLAB 2021a program was installed on Windows 11 and used to code the suggested technique. Training the suggested Parallel Attention–Residual CNN (PARC) model was carried out for multi-class and binary classification at the first step of the suggested methodology. The epoch value was chosen as 100 to achieve maximum performance. This ensured sufficient training iterations. The mini-batch size was chosen as 32. This was the maximum allowed by the hardware for the PARC model used. The initial learning rate was chosen as 0.001.

For smaller values, the training time increased and for larger values, the classification performance decreased. The SGDM technique was used as the optimization solver. The validation method employed was the 10-fold cross-validation, and the loss function was the cross-entropy. It took around 2 h to train the PARC model with the available hardware.

The loss values and accuracy graphics during optimization are presented in Figures 7 and 8. The training–validation accuracy scores for the binary classification reached 100% and 95.38%, and the training–validation loss values reached 0.035 and 0.1, respectively. The training–validation accuracy scores for the multi-class classification reached 99.21% and 91.89%, and the training–validation loss values reached 0.15 and 0.12, respectively.

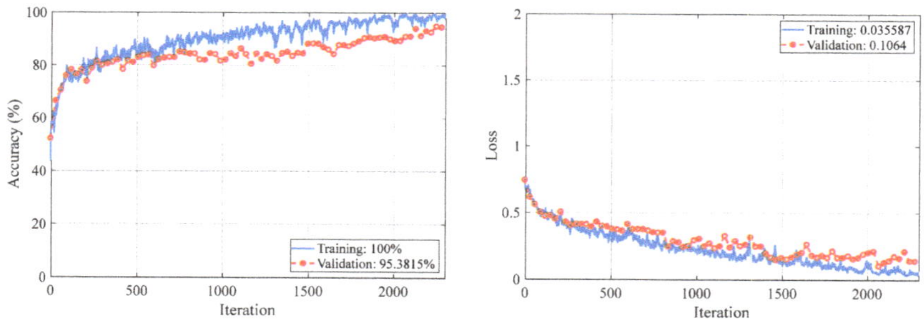

(a) (b)

Figure 7. Accuracy and loss graphs of the PARC model during the training process for binary classification: (**a**) accuracy graph, (**b**) loss graph.

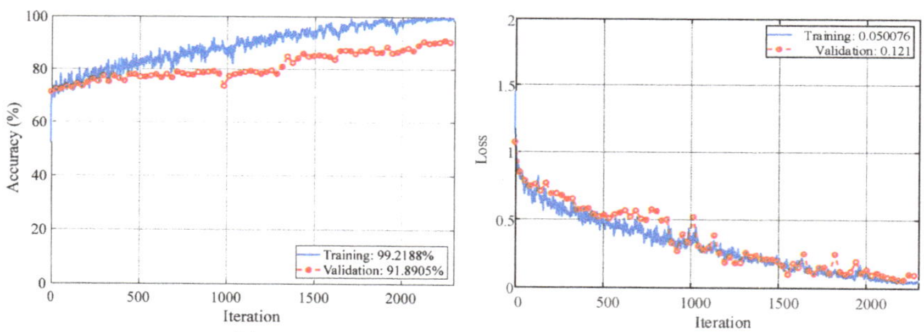

(a) (b)

Figure 8. Accuracy and loss graphs of the PARC model during the training process for multi–class classification: (**a**) accuracy graph, (**b**) loss graph.

By using learnable parameters and input data, the extraction of five hundred deep features was performed by "fc4", which is a fully connected layer in PARC. Therefore, the SVM algorithm could be used for the classification task instead of the softmax classifier. The Iterative Neighborhood Component Analysis (INCA) algorithm selected the most distinctive features and the computational cost decreased the execution time of the SVM classifier code. The number of the nearest neighbor (hyperparameter) was chosen as 10 (default value). Figure 9 shows the computed feature weights for each feature index. The threshold weights for the features to be selected with the INCA algorithm are set to 0.5 and 0.2 for binary and multi-class classification, respectively.

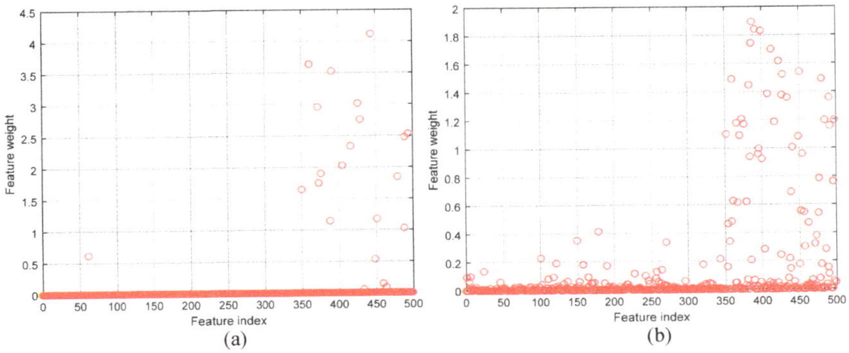

Figure 9. Features weighted with the INCA algorithm: (**a**) binary classification and (**b**) multi-class classification.

For binary and multi-class classification problems, 20 and 59 distinctive features were automatically selected by the INCA algorithm. Three-dimensional representations of the selected features are given in Figures 10 and 11 for two classification problems. In Figures 10 and 11, it is seen that the distinguishing characteristics of the features are increased with the feature selection process. In both figures, the rows represent classes and the columns show whether the feature selection process has been performed. The x-direction in the figures indicates the number of features. The y-direction provides the features' amplitude values while the z-direction represents the feature depth. To show the features in three dimensions, this parameter has been added. For each display, it is set to 2.

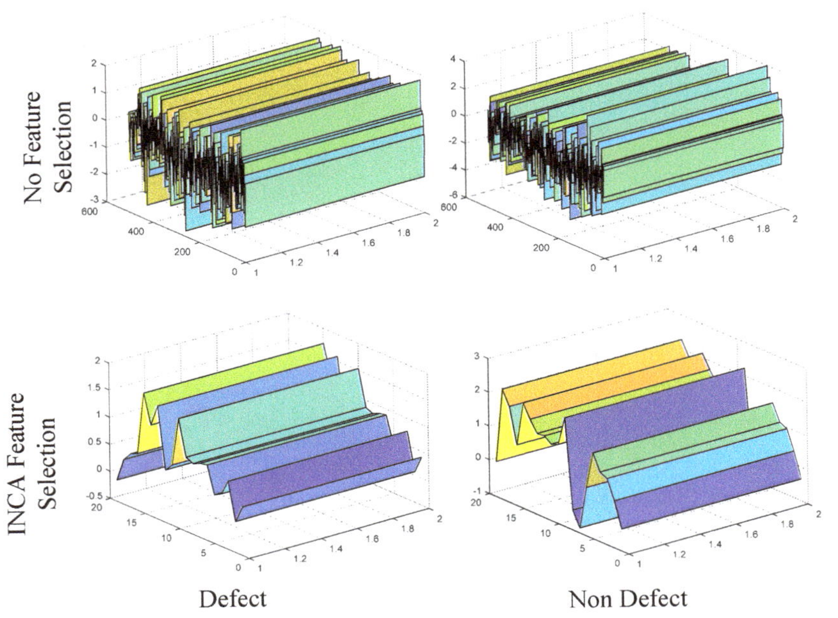

Figure 10. Three-dimensional feature representation for no feature selection and INCA feature selection cases (binary classification).

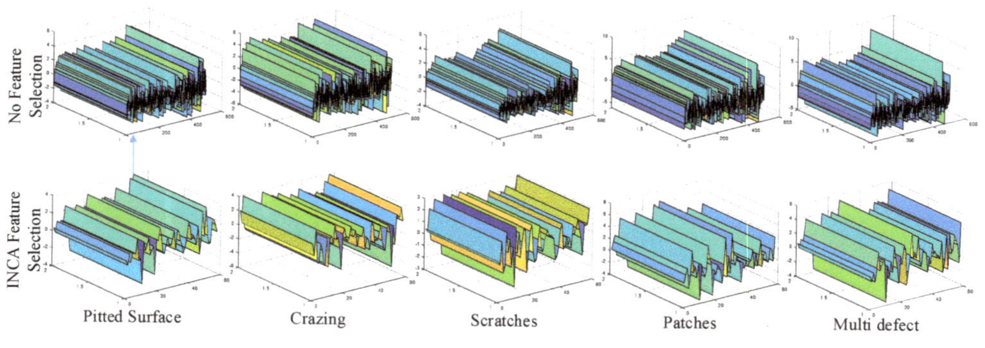

Figure 11. Three–dimensional feature representation for no feature selection and INCA feature selection cases (multi–class classification).

The accuracy scores according to feature selection cases and different classifiers are given in Table 1 for two classification problems. DT, SVM, KNN, NB, LD, Subspace KNN (SK), Subspace Discriminate (SD), and RUSBoosted Trees (RT) were classifier algorithms in the Classification Learner (CL) tool in the Matlab program. The default hyperparameters were selected in the CL tool for the classification process. This ablation study was performed to evaluate the feature selection operation's effectiveness and detect which classifier algorithm gave the best accuracy.

Table 1. Accuracy variations according to feature selection cases and different classifiers.

Classifier	No Feature Selection		INCA Feature Selection	
	Binary	Multi–Class	Binary	Multi–Class
DT	90.2	88.7	91.3	90
SVM	95.4	94.6	98.3	97.5
KNN	92.5	93.1	95.1	94.9
NB	93.2	87.9	95.1	90
LD	92.8	93.5	94.1	95,2
SK	90.1	92.8	92.3	95.6
SD	88.7	93.6	90.1	95.4
RT	89.3	85.6	91.8	88.5

As seen in Table 1, the SVM with the Gaussian kernel provided the best accuracy for two classification problems. In the case without feature selection, for binary and multi-class classification problems, the accuracy scores were 95.4% and 94.6%, respectively. In the case of the INCA feature selection, for binary and multi-class classification problems, the accuracy scores were 98.3% and 97.5%, respectively. Subspace Discriminant and RUSBoosted Trees provided the worst accuracies for two classification problems. In multi-class classification and binary classification without feature selection, the worst performance of the accuracy rate was 88.7% (Subspace Discriminant) and 85.6% (RUSBoosted Trees), respectively. In multi-class classification and binary classification with the INCA feature selection, the worst performances of the accuracy rate were 90.1% (Subspace Discriminant) and 88.5% (RUSBoosted Trees), respectively.

Figure 12 shows the confusion matrices of the proposed approach for the binary and multi-class classification tasks. In Figure 12a, the classes named 1, 2, 3, 4, and 5 represent the pitted surface, crazing, scratches, patches, and multi-class defect classes, respectively. In Figure 12b, the classes named 1, 2, 3, 4, and 5 represent the defect and non-defect classes, respectively. For the binary and multi-class classification problems, the accuracy scores were 98.3% and 97.5%, respectively.

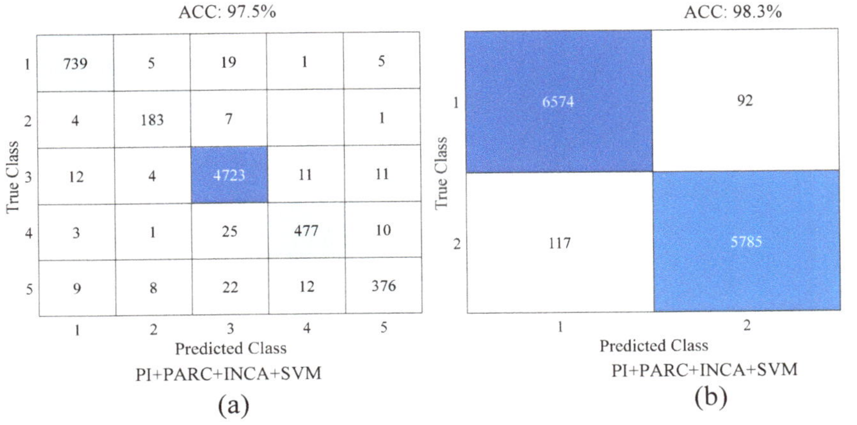

Figure 12. Confusion matrices of the proposed approach: (**a**) multi-class classification (classes: (1) pitted surfaces, (2) crazing, (3) scratches, (4) patches, (5) multi-class defect), (**b**) binary classification (1: defect 2: non-defect).

Performance metrics, including the sensitivity (SN), specificity (SP), precision (PR), and F-score, were computed by using true positive (TP), true negative (TN), false positive (FP), and false negative (FN) values. The results of the computed metrics are given in Table 2.

Table 2. Other performance metric results of the proposed approach for binary and multi-class classification.

Classification Mode	Classes	SN	SP	PR	F-Score
Binary	Defect	0.986	0.98	0.983	0.984
	Non-defect	0.98	0.986	0.984	0.982
Multi-Class	Pitted surfaces	0.961	0.995	0.963	0.962
	Crazing	0.938	0.997	0.91	0.924
	Scratches	0.992	0.96	0.985	0.988
	Patches	0.924	0.996	0.952	0.938
	Multi-class defects	0.881	0.996	0.933	0.906

For the binary classification, the SN, SP, PR, and F-score values of the defect class were 0.986, 0.98, 0.983, and 0.984, respectively. The SN, SP, PR, and F-score values of the non-defect class were 0.98, 0.986, 0.984, and 0.982, respectively. For the multi-class classification, the best SN (0.992), SP (0.996), PR (0.985), and F-score (0.988) values were obtained for the scratches, patches and multi-class defects, scratches, and scratches classes, respectively. The worst SN (0.881), SP (0.96), PR (0.985), and F-scores (0.988) values were obtained for the multi-class defects, scratches, crazing, and multi-class defect classes, respectively.

Figures 13 and 14 show the ROC curves and AUC values for the two classification problems. In Figure 13, positive classes 1 and 2 include the defect and non-defect classes, respectively. In Figure 14, positive classes 1, 2, 3, 4, and 5 include pitted surfaces, crazing, scratches, patches, and the multi-class defect classes, respectively.

As seen in Figure 13a,b, the AUC values were 0.99 for positive classes 1 and 2. As seen in Figure 14a–e, the AUC values were 0.99, 1.00, 0.99, 0.99, and 0.98 for positive classes 1, 2, 3, 4, and 5, respectively.

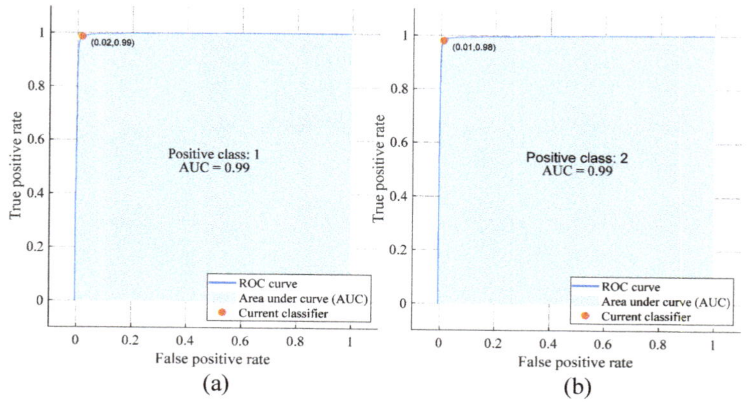

Figure 13. ROC curves and AUC values of the proposed approach: binary classification ((**a**) class 1: defect, (**b**) class 2: non-defect).

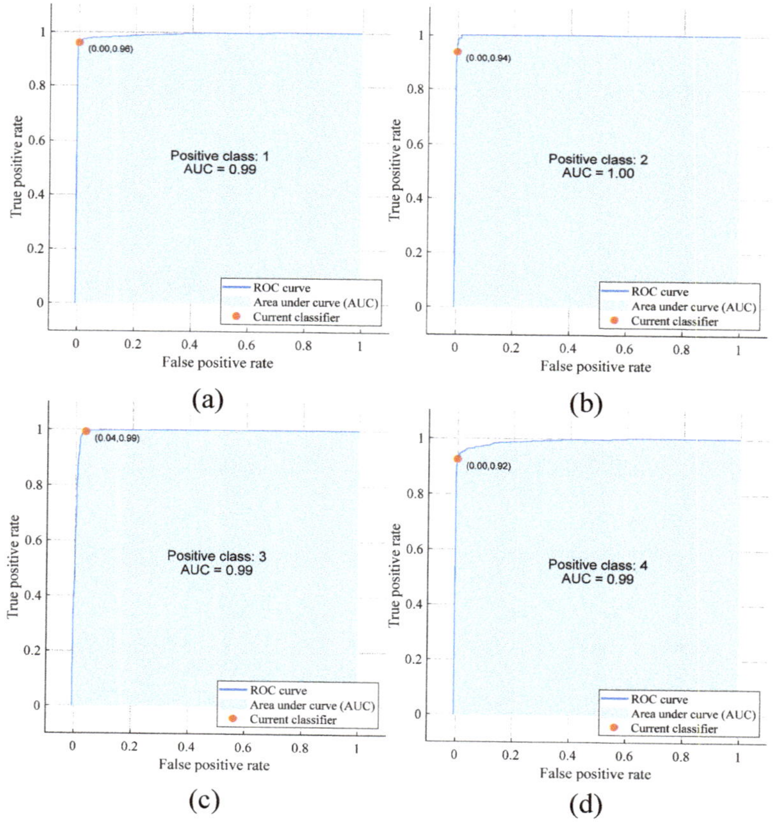

Figure 14. *Cont.*

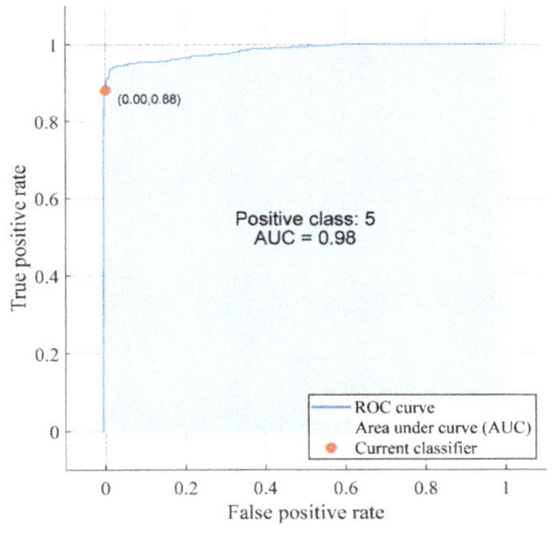

(e)

Figure 14. ROC curves and AUC values of the proposed approach: multi-class classification (classes: (**a**) pitted surfaces, (**b**) crazing, (**c**) scratches, (**d**) patches, (**e**) multi-class defects).

9. Discussion

In this section, ablation studies for the proposed approach were performed and the proposed approach was compared with state-of-the-art methods. Figures 15 and 16 show the effect of processed images (PI) and residual attention strategies on classification accuracy. Descriptive summary information about these ablation studies is provided in Table 3 for binary and multi-class classification, respectively.

Table 3. Summarizing information about the ablation study for the proposed approach ((**a**) binary classification, (**b**) multi-class classification).

Model Name	Model Info	Acc (a)	Acc (b)
RI + CNN	Raw images + PARC model without attention and residual structures	93.00%	87.20%
PI + CNN	Processed images + PARC model without attention and residual structures	93.50%	89.20%
PI + A-CNN	Processed images + PARC model without residual structures	93.90%	89.90%
PI + R-CNN	Processed images + PARC model without attention structures	94.50%	91.20%
PI + PARC	Processed images + PARC model (no feature selection with INCA and no SVM classifier)	95.40%	91.90%
PI + PARC + SVM	Processed images + PARC model +SVM (no feature selection with INCA)	96.80%	93.90%
PI + PARC + INCA + SVM	Proposed approach	98.30%	97.50%

As seen in Figure 15 for the binary classification, the best accuracy was obtained by the proposed approach (PI + PARC + SVM) while the worst accuracy was obtained by the raw image (RI) + CNN strategy (Figure 15a). In Figure 15b, PI instead of RI was used for the classification and the classification accuracy was improved by 3.5%. In Figure 15c, the attention structure (PI + A-CNN) was added to the CNN model in Figure 15b. The classification accuracy was improved by 0.4%. In Figure 15d, the residual structure

(PI + R-CNN) was added to the CNN model in Figure 15b. The classification accuracy was improved by 1.0%. In Figure 15e, the parallel residual and attention structure (PI + PARC) was added to the CNN model in Figure 15b. The classification accuracy was improved by 1.5%. In Figure 15f, the SVM classifier with INCA feature selection algorithm was applied in place of the softmax classifier in the proposed Parallel Attention–Residual CNN (PARC) model and the accuracy of classification was improved by 2.9% compared to the model in Figure 15e.

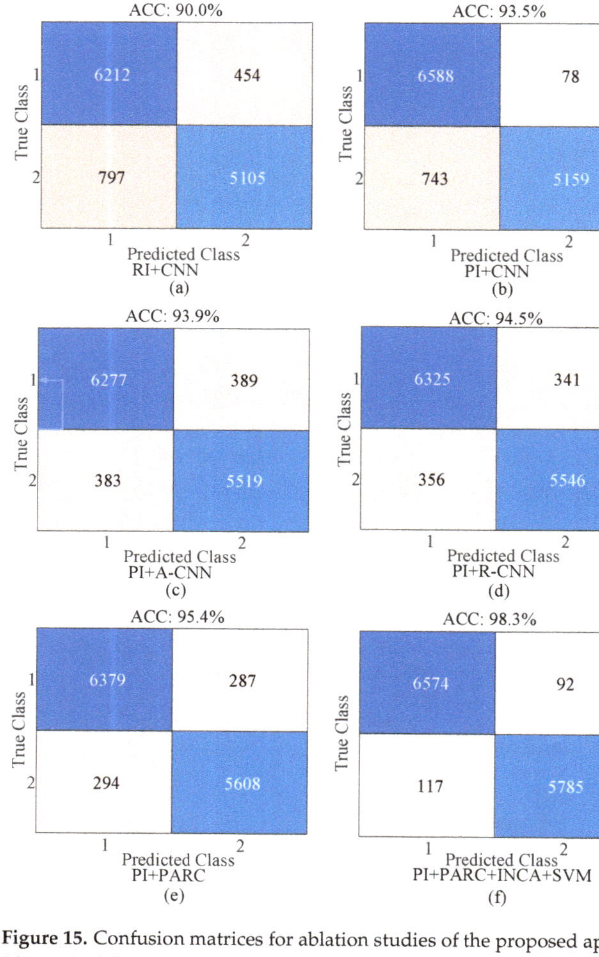

Figure 15. Confusion matrices for ablation studies of the proposed approach (binary classification (classes 1: defect 2: non-defect)).

As seen in Figure 16 for the multi-class classification, the best accuracy was obtained by the proposed approach (PI + PARC + SVM) while the worst accuracy was obtained by the raw image (RI) + CNN strategy (Figure 16a). In Figure 16b, PI instead of RI was used for the classification, and the accuracy performance of classification was increased by 2.0%. In Figure 16c, attention structure (PI + A-CNN) was added to the CNN model in Figure 16b. The classification accuracy was improved by 0.7%. In Figure 16d, the residual structure (PI + R-CNN) was added to the CNN model in Figure 16b. The classification accuracy was improved by 2.0%. In Figure 16e, the parallel residual and attention structure (PI + PARC) was added to the CNN model in Figure 16b. The classification accuracy was improved by 2.7%. In Figure 16f, the SVM classifier with the INCA algorithm was used instead of the softmax classifier in the suggested PARC method, and the accuracy performance of classification was improved by 5.6% in comparison with the model in Figure 16e.

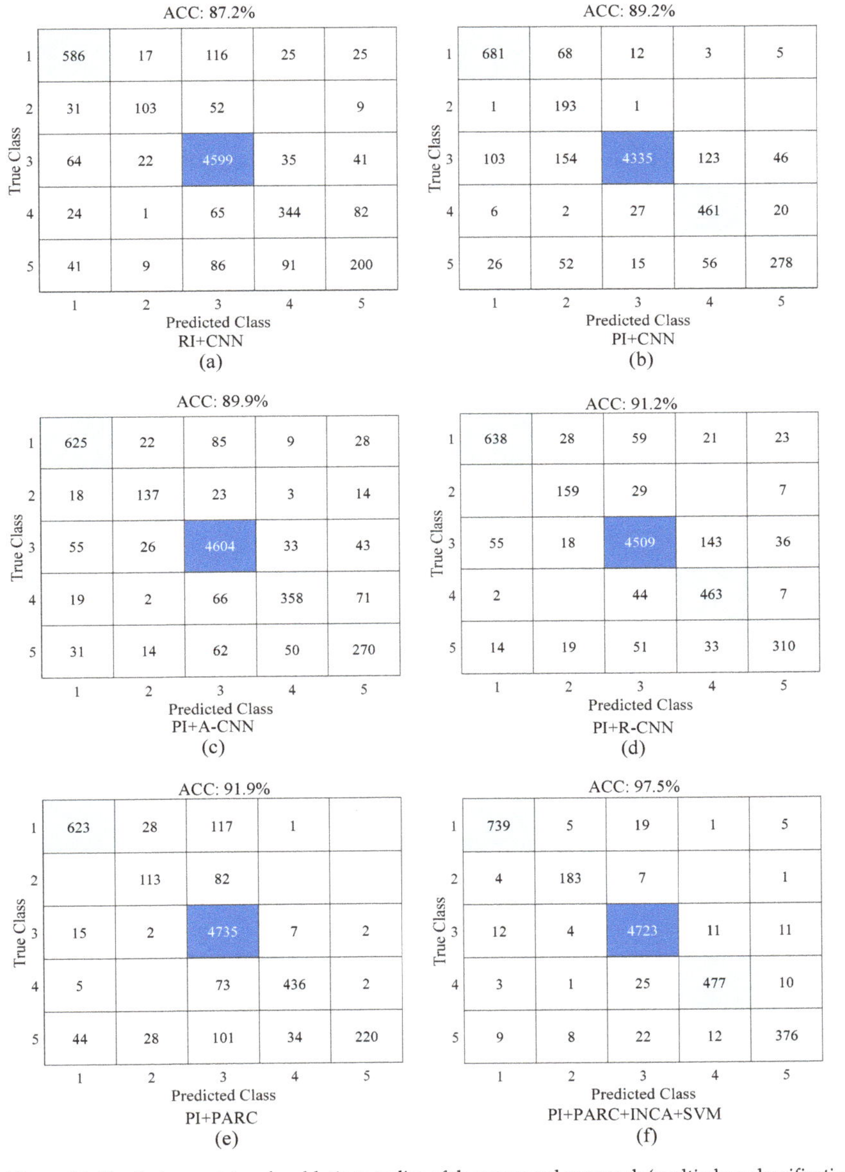

Figure 16. Confusion matrices for ablation studies of the proposed approach (multi-class classification (classes: (1) pitted surfaces, (2) crazing, (3) scratches, (4) patches, (5) multi-class defects)).

Table 4 presents the performance results of the proposed approach and state-of-the-art models using the dataset named "Severstal: Steel Defect Detection". Fadli and Herlistiono [50] used the Xception network, a pre-trained CNN model for automated steel surface classification. For the binary and multi-class classification, the classification accuracies were

94% and 85%, respectively. Guo et al. [51] performed automated surface steel detection with a specific GAN model. The binary class classification performance reached 96.80% accuracy. A hybrid approach with Faster R-CNN models and ResNet50 was proposed by Wang et al. [21]. The prediction values obtained from weight activations of the ResNet model were utilized for the thresholding operation. If the scores were less than 0.3, the steel samples were considered to be defect-free. If scores were larger than 0.3, the samples were described as defective. With this method, the binary classification accuracy was 97.47%. Additionally, four copies of each steel sample image from the dataset were created, along with new class labels, using this method. The dataset was therefore multiplied by four. Steel surface flaw classification was conducted by Chigateri et al. [52] using the Exception model. They achieved an accuracy of 88% for binary classification and the same accuracy of 88% for multiple classification. A combination of the ResNet model with the squeeze-and-excitation networks suggested by Hu et al. [53] produced an accuracy value of 87.5% for the two-class classification task and 94% for the four-class classification problem.

Table 4. The classification accuracies of studies and the proposed model using the same dataset.

Authors	Methodology	Accuracy (%)	
		Binary	Multi-Class
Proposed model	PI + PARC + INCA + SVM	98.9	94.5
Wang et al. [21]	ResNet50 + Faster R-CNN	97.47	-
Guo et al. [51]	GAN model	96.8	-
Chigateri et al. [52]	Exception model	87.6	85.0
Hu et al. [53]	SEResNET50	94.0	87.5
Fadli et al. [50]	Transfer learning (Xception)	94.0	85.0

The efficacy of the proposed method was assessed by employing an alternate dataset, specifically the NEU Surface Defect Database [54]. Consequently, the dependability of the suggested approach was enhanced. This dataset comprised six distinct types of surface defects: crazing (class 1), inclusions (class 2), patches (class 3), pitted surfaces (class 4), rolled-in scales (class 5), and scratches (class 6). Through the utilization of 10-fold cross-validation, the dataset, consisting of a total of 1800 samples distributed evenly with 300 examples in each class, underwent evaluation. Within this dataset, the suggested technique demonstrated a classification accuracy of 99.77%, as depicted in the confusion matrix presented in Figure 17.

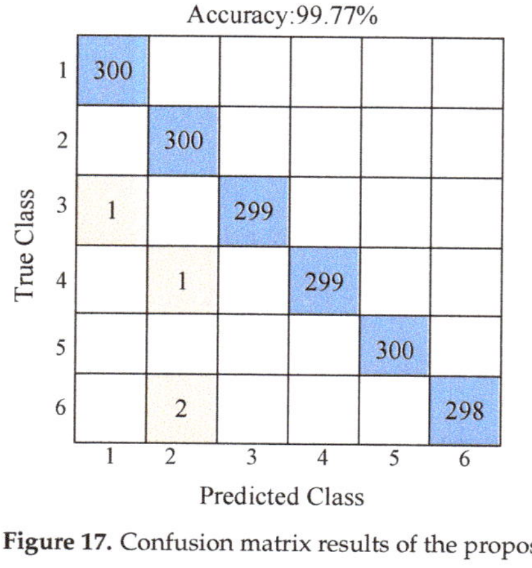

Figure 17. Confusion matrix results of the proposed approach on the NEU Surface Defect Database.

As can be seen from the results in Tables 4 and 5, the proposed model improved the classification performance compared to other models using the same dataset. However, given that the methods' training parameters and training–evaluation methodology differ, this finding does not necessarily imply that the proposed strategy is better than the others.

Table 5. The classification accuracies of studies and the proposed model using the NEU Surface Defect Database.

Author	Method	Accuracy (%)
Yeung et al. [54]	Fused Attention CNN model	89.3
Tian et al. [55]	SegNet + CNN	89.6
Yi et al. [56]	Deep CNN	99.05
Li et al. [14]	Transfer learning (ResNet)	99.0
Fu et al. [15]	Lightweight CNN	99.61
Proposed approach	PARC, NRMI, SVM	99.77

Table 5 presents the performance results of the proposed approach and state-of-the-art models on the NEU Surface Defect Database. Yeung et al. [54] introduced a novel hybrid model, which underwent training from the ground up by incorporating the CNN architecture and five attention layers. The resultant CNN model, enriched with integrated attention, demonstrated an accuracy of 89.30%. Meanwhile, Tian et al. [55] employed the SegNet model to reconstruct and segment images of steel surfaces, and subsequent classification using a CNN model with an end-to-end learning approach yielded a success rate of 89.60%. Yi et al. [56] utilized a 14-layer CNN model featuring five convolutional layers, achieving an impressive classification accuracy of 99.05%. Li et al. [14] implemented a transfer learning system based on the ResNet model, attaining a classification accuracy of 99.00% with the ResNet CNN model. In another approach, Fu et al. [15] leveraged the SqueezeNet framework to construct a lightweight CNN model, incorporating a blur operation for raw photo processing, resulting in a remarkable success percentage of 99.61%.

The proposed approach outperformed the CNN, pre-trained CNN, R-CNN, transfer learning, and GAN models due to its novel and task-specific design. Key innovations included the use of a spectrogram algorithm in pre-processing, which transformed image data into a more detailed representation of surface defects, and the Parallel Attention–Residual CNN (PARC) model, which combined attention mechanisms for highlighting critical regions and residual connections for preserving important feature information. Unlike traditional models that relied on softmax for classification, this approach extracted deep features and evaluated them using multiple classifiers, with the SVM yielding the best results. Additionally, the INCA feature selection algorithm reduced the computational complexity while improving the classification accuracy. This tailored methodology enhanced defect detection by focusing on capturing subtle pixel variations and optimizing performance for the specific task.

Commonly used CNN models in both datasets include either pre-trained CNN models, such as ResNet, or lightly weighted CNN models. The network file sizes of the proposed PARC model and other popular pre-trained models with weights are given in Table A2. The PARC model has less weight than the other pre-trained CNN models, except the MobileNet model. Therefore, the execution time is optimal for both training and testing. In addition, accurate classification is more important than speed in the detection of steel surface defects.

Table 6 presents the classification accuracy results for both multi-class and binary classification tasks using various image enhancement techniques followed by CNN-based classification. The proposed spectrogram-based method shows superior performance in both scenarios.

Table 6. Comparison of the proposed pre-processing method with basic image enhancement algorithms.

Method	Description	Multi-Class Accuracy (%)	Binary Accuracy (%)
Raw Images + CNN	Baseline method without any pre-processing.	93	87.2
Histogram Equalization + CNN	Enhances global contrast by redistributing image intensities.	93.4	88.7
Gamma Correction	Adjusts brightness using nonlinear intensity mapping.	93.2	87.9
CLAHE + CNN	Applies adaptive histogram equalization locally to improve contrast in homogeneous areas.	93.3	88.9
Unsharp Masking	Sharpens the image by emphasizing edges and details.	93	87.5
Log/Power-Law Transform	Enhances low-intensity pixels, useful for improving contrast in dark regions.	93.2	88.6

The experimental results demonstrate the effectiveness of the proposed spectrogram-based method in enhancing the discriminability of surface defects for both multi-class and binary classification tasks. While the baseline approach using raw images with CNN achieved a 93.0% accuracy for multi-class and 87.2% for binary classification, applying conventional image enhancement techniques such as histogram equalization, gamma correction, CLAHE, unsharp masking, and log/power-law transforms led to modest improvements. Among these, CLAHE and histogram equalization performed relatively better due to their ability to improve the local and global contrast, respectively. However, the proposed method outperformed all the others, achieving the highest classification accuracy of 93.5% in the multi-class scenario and 89.2% in binary classification.

This superior performance can be attributed to the spectrogram's ability to transform gradient-derived signals into the time–frequency domain, enabling the capture of both spatial patterns and frequency-based features related to surface defects. Unlike conventional techniques that primarily operate in the spatial domain and focus on intensity or contrast, the spectrogram representation provides a richer and more informative input to the CNN by highlighting subtle defect structures. These results quantitatively validate the advantage of the proposed method over traditional enhancement techniques in terms of feature extraction and defect classification performance.

10. Conclusions

This study focuses on the automatic detection of surface defects, which is an important issue in steel fabrication. It has been made to increase the automatic classification performance with a specific deep learning-based strategy. The classification performance is enhanced by processed images and feature extraction in the Parallel Attention–Residual CNN (PARC) model. The PARC model outperformed the CNN model (no residual and attention structures), the R-CNN (Residual-CNN), and the A-CNN (Attention-CNN) models. In addition, the Iterative Neighborhood Component Analysis (INCA) algorithm efficiently reduced the size of the feature set and improved the classification performance for both datasets. The classification performance values obtained for the Severstal Dataset (SD) and the Neu-Surface Dataset (NSD) are summarized in Table 7. As can be seen from Table 7, both datasets achieved 0.94 and above in all the performance metrics. In both datasets, the classification accuracy was improved according to the model that achieves the best performance out of the existing models. The classification performance is improved by 1.43%, 7.0%, and 0.16% for SD 2-class, SD 4-class, and NSD, respectively.

Table 7. Summary of the experimental results of the proposed method.

Criteria	SD 2-Class	SD 4-Class	NSD
Accuracy (%)	98.9	94.5	99.7
Sensitivity	0.98	0.94	0.99
Specificity	0.98	0.98	0.99
Precision	0.98	0.94	0.98
F-score	0.98	0.94	0.98
Performance improvement (%)	1.43	7.0	0.16

The good classification performance of the proposed approach makes it possible to use it for the real-time detection of steel surface defects. In the next phase, the recorded weights of the proposed approach can be tested on an artificial intelligence development kit such as an NVIDIA Jetson Orin Nano (NVIDIA, Santa Clara, CA, USA). Thus, once the test performance of the model is confirmed, it can be used in enterprises.

The most important limitation of the proposed model is that it is difficult to implement in real-time embedded systems due to the size of the model. This limitation can be solved with server-based systems. However, this may increase the financial cost.

Author Contributions: Conceptualization, F.D. and K.S.P.; methodology, F.D.; software, F.D.; validation, F.D. and K.S.P.; formal analysis, F.D. and K.S.P.; investigation, F.D. and K.S.P.; resources, F.D. and K.S.P.; data curation, F.D. and K.S.P.; writing—original draft preparation, F.D. and K.S.P.; writing—review and editing, F.D. and K.S.P.; visualization, F.D. and K.S.P.; supervision, F.D. and K.S.P.; project administration, F.D. and K.S.P.; funding acquisition, F.D. and K.S.P. All authors have read and agreed to the published version of the manuscript.

Funding: This research received no external funding.

Institutional Review Board Statement: Not applicable.

Informed Consent Statement: Not applicable.

Data Availability Statement: The datasets are publicly available on https://github.com/siddhartamukherjee/NEU-DET-Steel-Surface-Defect-Detection (accessed on 10 March 2025), https://www.kaggle.com/competitions/severstal-steel-defect-detection (accessed on 10 March 2025).

Conflicts of Interest: The authors declare no conflicts of interest.

Appendix A

Table A1. Layers of the PARC model.

Layer	Layer Order	Layer	Layer Info
1	'conv2d_1'	2D Convolution	In total, 16 5 × 5 × 3 convolutions with stride [1 1] and padding 'same'
2	'batchnorm_1'	Batch Normalization	Batch normalization with 16 channels
3	'relu1'	ReLU	ReLU
4	'maxpool2d_1'	2D Max Pooling	In total, 2 × 2 max pooling with stride [1 1] and padding [0 0 0 0]
5	'conv2d2_1'	2D Convolution	In total, 8 3 × 3 × 16 convolutions with stride [1 1] and padding 'same'

Table A1. *Cont.*

Layer	Layer Order	Layer	Layer Info
6	'batchnorm_2'	Batch Normalization	Batch normalization with 8 channels
7	'relu2'	ReLU	ReLU
8	'maxpool2d2'	2D Max Pooling	In total, 2 × 2 max pooling with stride [1 1] and padding [0 0 0 0]
9	'conv2d_2'	2D Convolution	In total, 8 2 × 2 × 8 convolutions with stride [1 1] and padding 'same'
10	'relu_2_1_1_5'	ReLU	ReLU
11	'conv2d_3'	2D Convolution	In total, 8 3 × 3 × 8 convolutions with stride [1 1] and padding 'same'
12	'maxpool2d_2'	2D Max Pooling	In total, 2 × 2 max pooling with stride [1 1] and padding 'same'
13	'conv2d2_2'	2D Convolution	In total, 8 5 × 5 × 8 convolutions with stride [1 1] and padding 'same'
14	'relu_2_1_1_6'	ReLU	ReLU
15	'conv2d2_3'	2D Convolution	In total, 8 3 × 3 × 8 convolutions with stride [1 1] and padding 'same'
16	'sigmoid_1_1_1_3'	sigmoidLayer	sigmoidLayer
17	'mul_1_1_1_3'	ElementWiseMultiplication	Element-Wise Multiplication of 2 inputs
18	'fc0'	Fully Connected	In total, 350 fully connected layers
19	'ReLu2'	ReLU	ReLU
20	'input'	Image Input	In total, 32 × 200 × 3 images with 'zero center' normalization
21	'convInp'	2D Convolution	In total, 16 3 × 3 × 3 convolutions with stride [1 1] and padding 'same'
22	'BNInp'	Batch Normalization	Batch normalization with 16 channels
23	'reluInp'	ReLU	ReLU
24	'S1U1conv1'	2D Convolution	In total, 16 3 × 3 × 16 convolutions with stride [1 1] and padding 'same'
25	'S1U1BN1'	Batch Normalization	Batch normalization with 16 channels
26	'S1U1relu1'	ReLU	ReLU
27	'S1U1conv2'	2D Convolution	In total, 16 3 × 3 × 16 convolutions with stride [1 1] and padding 'same'
28	'S1U1BN2'	Batch Normalization	Batch normalization with 16 channels
29	'add11'	Addition	Element-wise addition of 2 inputs
30	'relu11'	ReLU	ReLU
31	'S1U2conv1'	2D Convolution	In total, 16 3 × 3 × 16 convolutions with stride [1 1] and padding 'same'
32	'S1U2BN1'	Batch Normalization	Batch normalization with 16 channels
33	'S1U2relu1'	ReLU	ReLU
34	'S1U2conv2'	2D Convolution	In total, 16 3 × 3 × 16 convolutions with stride [1 1] and padding 'same'
35	'S1U2BN2'	Batch Normalization	Batch normalization with 16 channels
36	'add12'	Addition	Element-wise addition of 2 inputs
37	'relu12'	ReLU	ReLU
38	'S2U1conv1'	2D Convolution	In total, 32 3 × 3 × 16 convolutions with stride [2 2] and padding 'same'
39	'S2U1BN1'	Batch Normalization	Batch normalization with 32 channels
40	'S2U1relu1'	ReLU	ReLU
41	'S2U1conv2'	2D Convolution	In total, 32 3 × 3 × 32 convolutions with stride [1 1] and padding 'same'
42	'S2U1BN2'	Batch Normalization	Batch normalization with 32 channels
43	'add21'	Addition	Element-wise addition of 2 inputs
44	'relu21'	ReLU	ReLU

Table A1. *Cont.*

Layer	Layer Order	Layer	Layer Info
45	'S2U2conv1'	2D Convolution	In total, 32 3 × 3 × 32 convolutions with stride [1 1] and padding 'same'
46	'S2U2BN1'	Batch Normalization	Batch normalization with 32 channels
47	'S2U2relu1'	ReLU	ReLU
48	'S2U2conv2'	2D Convolution	In total, 32 3 × 3 × 32 convolutions with stride [1 1] and padding 'same'
49	'S2U2BN2'	Batch Normalization	Batch normalization with 32 channels
50	'add22'	Addition	Element-wise addition of 2 inputs
51	'relu22'	ReLU	ReLU
52	'S3U1conv1'	2D Convolution	In total, 64 3 × 3 × 32 convolutions with stride [2 2] and padding 'same'
53	'S3U1BN1'	Batch Normalization	Batch normalization with 64 channels
54	'S3U1relu1'	ReLU	ReLU
55	'S3U1conv2'	2D Convolution	In total, 64 3 × 3 × 64 convolutions with stride [1 1] and padding 'same'
56	'S3U1BN2'	Batch Normalization	Batch normalization with 64 channels
57	'add31'	Addition	Element-wise addition of 2 inputs
58	'relu31'	ReLU	ReLU
59	'S3U2conv1'	2D Convolution	In total, 64 3 × 3 × 64 convolutions with stride [1 1] and padding 'same'
60	'S3U2BN1'	Batch Normalization	Batch normalization with 64 channels
61	'S3U2relu1'	ReLU	ReLU
62	'S3U2conv2'	2D Convolution	In total, 64 3 × 3 × 64 convolutions with stride [1 1] and padding 'same'
63	'S3U2BN2'	Batch Normalization	Batch normalization with 64 channels
64	'add32'	Addition	Element-wise addition of 2 inputs
65	'relu32'	ReLU	ReLU
66	'globalPool'	2D Average Pooling	In total, 8 × 8 average pooling with stride [1 1] and padding [0 0 0 0]
67	'fc3'	Fully Connected	In total, 350 fully connected layers
68	'relu_3'	ReLU	ReLU
69	'add42'	Addition	Element-wise addition of 2 inputs
70	'fc4'	Fully Connected	In total, 150 fully connected layers
71	'relu_4'	ReLU	ReLU
72	'fcFinal'	Fully Connected	In total, 2 fully connected layers
73	'softmax'	Softmax	softmax
74	'classoutput'	Classification Output	crossentropyex with classes 'defect' and 'non_defect'
75	'skipConv1'	2D Convolution	In total, 32 1 × 1 × 16 convolutions with stride [2 2] and padding [0 0 0 0]
76	'skipBN1'	Batch Normalization	Batch normalization with 32 channels
77	'skipConv2'	2D Convolution	In total, 64 1 × 1 × 32 convolutions with stride [2 2] and padding [0 0 0 0]
78	'skipBN2'	Batch Normalization	Batch normalization with 64 channels

Table A2. Sizes of the PARC model and popular pre-trained CNN models.

CNN Model	Model File Size (MB)
ResNet	93.2
VGGNet	502.9
AlexNet	222.03
MobileNet	12.95
PARC Model	64.7

Table A3. Pseudocode for the implementation steps of the proposed method.

1. Begin with raw images containing surface defects.
2. Stage 1: Pre-processing
    a. Apply spectrogram algorithm to the time series signals of the raw images.
    b. Convert 1D data using all pixel values from the images.
    c. Obtain spectrogram images representing pixel changes in defect regions.
3. Stage 2: Model Training with PARC (Parallel Attention–Residual CNN)
    a. Design and initialize the PARC model.
        i. Combine attention and residual modules with a customized CNN.
        ii. Train attention module to highlight important image regions.
        iii. Use residual module to pass feature maps from earlier layers to later convolutional layers.
    b. Train the PARC model with the processed spectrogram images.
4. Stage 3: Feature Extraction
    a. Extract deep features from the fully connected (FC) layer of the PARC model.
    b. Use these features (instead of the softmax classifier) to test with other classifier algorithms.
5. Stage 4: Feature Selection
    a. Apply the INCA feature selection algorithm.
    b. Reduce computational cost and improve classification performance by selecting the most relevant features.
6. Stage 5: Classification
    a. Train seven popular classifiers on the selected features:
        i. Decision Tree (DT).
        ii. Support Vector Machine (SVM).
        iii. K-Nearest Neighbor (KNN).
        iv. Naïve Bayes (NB).
        v. Linear Discriminant (LD).
        vi. Subspace KNN.
        vii. Subspace Discriminate.
        viii. RUSBoosted Trees.
    b. Evaluate the classification performance of each algorithm.
    c. Select the best-performing algorithm (SVM) for the final classification.
7. End

References

1. Zhang, D.; Song, K.; Xu, J.; He, Y.; Niu, M.; Yan, Y. MCnet: Multiple Context Information Segmentation Network of No-Service Rail Surface Defects. *IEEE Trans. Instrum. Meas.* **2021**, *70*, 5004309. [CrossRef]
2. Hanbay, K.; Talu, M.F.; Özgüven, Ö.F. Fabric Defect Detection Systems and Methods—A Systematic Literature Review. *Optik* **2016**, *127*, 11960–11973. [CrossRef]
3. Zhao, W.; Chen, F.; Huang, H.; Li, D.; Cheng, W. A New Steel Defect Detection Algorithm Based on Deep Learning. *Comput. Intell. Neurosci.* **2021**, *2021*, 5592878. [CrossRef] [PubMed]
4. Dong, H.; Song, K.; He, Y.; Xu, J.; Yan, Y.; Meng, Q. PGA-Net: Pyramid Feature Fusion and Global Context Attention Network for Automated Surface Defect Detection. *IEEE Trans. Ind. Inform.* **2020**, *16*, 7448–7458. [CrossRef]
5. Cao, J.; Yang, G.; Yang, X. A Pixel-Level Segmentation Convolutional Neural Network Based on Deep Feature Fusion for Surface Defect Detection. *IEEE Trans. Instrum. Meas.* **2021**, *70*, 5003712. [CrossRef]
6. Dikbas, H.; Taskaya, S. Alloying the Surface of AISI 2205 Duplex Stainless Steel Material by PTA Welding Method and Making Its Thermomechanical Investigation in ANSYS Software. *J. Therm. Anal. Calorim.* **2020**, *139*, 3847–3856. [CrossRef]
7. Konovalenko, I.; Maruschak, P.; Brezinová, J.; Viňáš, J.; Brezina, J. Steel Surface Defect Classification Using Deep Residual Neural Network. *Metals* **2020**, *10*, 846. [CrossRef]
8. den Bakker, I. *Python Deep Learning Cookbook: Over 75 Practical Recipes on Neural Network Modeling, Reinforcement Learning, and Transfer Learning Using Python*; Packt Publishing Ltd.: Birmingham, UK, 2007; Volume 136, ISBN 9781787125193.
9. Martins, L.A.O.; Pádua, F.L.C.; Almeida, F.E.M. Automatic Detection of Surface Defects on Rolled Steel Using Computer Vision and Artificial Neural Networks. In *Proceedings of the IECON Proceedings (Industrial Electronics Conference), Glendale, Arizona, USA, 7–10 November 2010*; IEEE: New York, NY, USA, 2010; pp. 1081–1086.

10. Li, Z.; Wei, X.; Hassaballah, M.; Li, Y.; Jiang, X. A Deep Learning Model for Steel Surface Defect Detection. *Complex Intell. Syst.* **2024**, *10*, 885–897. [CrossRef]
11. Pang, W.; Tan, Z. A Steel Surface Defect Detection Model Based on Graph Neural Networks. *Meas. Sci. Technol.* **2024**, *35*, 46201. [CrossRef]
12. Zhang, H.; Li, S.; Miao, Q.; Fang, R.; Xue, S.; Hu, Q.; Hu, J.; Chan, S. Surface Defect Detection of Hot Rolled Steel Based on Multi-Scale Feature Fusion and Attention Mechanism Residual Block. *Sci. Rep.* **2024**, *14*, 7671. [CrossRef]
13. Zhao, Y.J.; Yan, Y.H.; Song, K.C. Vision-Based Automatic Detection of Steel Surface Defects in the Cold Rolling Process: Considering the Influence of Industrial Liquids and Surface Textures. *Int. J. Adv. Manuf. Technol.* **2017**, *90*, 1665–1678. [CrossRef]
14. Li, J.; Su, Z.; Geng, J.; Yin, Y. Real-Time Detection of Steel Strip Surface Defects Based on Improved YOLO Detection Network. *IFAC-PapersOnLine* **2018**, *51*, 76–81. [CrossRef]
15. Fu, G.; Sun, P.; Zhu, W.; Yang, J.; Cao, Y.; Yang, M.Y.; Cao, Y. A Deep-Learning-Based Approach for Fast and Robust Steel Surface Defects Classification. *Opt. Lasers Eng.* **2019**, *121*, 397–405. [CrossRef]
16. Liu, Y.; Xu, K.; Xu, J. Periodic Surface Defect Detection in Steel Plates Based on Deep Learning. *Appl. Sci.* **2019**, *9*, 3127. [CrossRef]
17. Liu, Y.; Yuan, Y.; Balta, C.; Liu, J. A Light-Weight Deep-Learning Model with Multi-Scale Features for Steel Surface Defect Classification. *Materials* **2020**, *13*, 4629. [CrossRef]
18. Guan, S.; Lei, M.; Lu, H. A Steel Surface Defect Recognition Algorithm Based on Improved Deep Learning Network Model Using Feature Visualization and Quality Evaluation. *IEEE Access* **2020**, *8*, 49885–49895. [CrossRef]
19. Amin, D.; Akhter, S. Deep Learning-Based Defect Detection System in Steel Sheet Surfaces. In *Proceedings of the 2020 IEEE Region 10 Symposium, TENSYMP 2020, Dhaka, Bangladesh, 5–7 June 2020*; IEEE: New York, NY, USA, 2020; pp. 444–448.
20. Severstal: Steel Defect Detection. Available online: https://www.kaggle.com/c/severstal-steel-defect-detection/overview (accessed on 10 March 2025).
21. Wang, S.; Xia, X.; Ye, L.; Yang, B. Automatic Detection and Classification of Steel Surface Defect Using Deep Convolutional Neural Networks. *Metals* **2021**, *11*, 388. [CrossRef]
22. Demir, F.; Bajaj, V.; Ince, M.C.; Taran, S.; Şengür, A. Surface EMG Signals and Deep Transfer Learning-Based Physical Action Classification. *Neural Comput. Appl.* **2019**, *31*, 8455–8462. [CrossRef]
23. Özseven, T. Investigation of the Effect of Spectrogram Images and Different Texture Analysis Methods on Speech Emotion Recognition. *Appl. Acoust.* **2018**, *142*, 70–77. [CrossRef]
24. Kollias, D.; Zafeiriou, S. Exploiting Multi-CNN Features in CNN-RNN Based Dimensional Emotion Recognition on the OMG in-the-Wild Dataset. *IEEE Trans. Affect. Comput.* **2021**, *12*, 595–606. [CrossRef]
25. Demir, K.; Berna, A.R.I.; Demir, F. Detection of Brain Tumor with a Pre-Trained Deep Learning Model Based on Feature Selection Using MR Images. *Firat Univ. J. Exp. Comput. Eng.* **2023**, *2*, 23–31. [CrossRef]
26. Canan, K.O.Ç.; Özyurt, F. An Examination of Synthetic Images Produced with DCGAN According to the Size of Data and Epoch. *Firat Univ. J. Exp. Comput. Eng.* **2023**, *2*, 32–37.
27. Krizhevsky, A.; Sutskever, I.; Hinton, G.E. Imagenet Classification with Deep Convolutional Neural Networks. In Proceedings of the Advances in Neural Information Processing Systems, Lake Tahoe, NV, USA, 3–6 December 2012; pp. 1097–1105.
28. Simonyan, K.; Zisserman, A. Very Deep Convolutional Networks for Large-Scale Image Recognition. *arXiv* **2014**, arXiv:1409.1556.
29. Ioffe, S.; Szegedy, C. Batch Normalization: Accelerating Deep Network Training by Reducing Internal Covariate Shift. In Proceedings of the 32nd International Conference on Machine Learning ICML 2015, Lille, France, 7–9 July 2015; Volume 1, pp. 448–456.
30. Santurkar, S.; Tsipras, D.; Ilyas, A.; Mądry, A. How Does Batch Normalization Help Optimization? In Proceedings of the 32nd International Conference on Neural Information Processing Systems, Montreal, QC, Canada, 2–8 December 2018; pp. 2488–2498.
31. Agarap, A.F. Deep Learning Using Rectified Linear Units (Relu). *arXiv* **2018**, arXiv:1803.08375.
32. Weng, L.; Zhang, H.; Chen, H.; Song, Z.; Hsieh, C.-J.; Daniel, L.; Boning, D.; Dhillon, I. Towards Fast Computation of Certified Robustness for Relu Networks. In Proceedings of the International Conference on Machine Learning, PMLR, Stockholm, Sweden, 10–15 July 2018; pp. 5276–5285.
33. Yilmaz, T.; Aydogmus, O. Deep Deterministic Policy Gradient Reinforcement Learning for Collision-Free Navigation of Mobile Robots in Unknown Environments. *Firat Univ. J. Exp. Comput. Eng.* **2023**, *2*, 87–96. [CrossRef]
34. Liang, X.; Wang, X.; Lei, Z.; Liao, S.; Li, S.Z. Soft-Margin Softmax for Deep Classification. In Proceedings of the International Conference on Neural Information Processing, Guangzhou, China, 14–18 November 2017; pp. 413–421.
35. Zang, F.; Zhang, J. Softmax Discriminant Classifier. In Proceedings of the 2011 Third International Conference on Multimedia Information Networking and Security, Shanghai, China, 4–6 November 2011; pp. 16–19.
36. Akpinar, E.K.; Mehmet, D.A.Ş. Modeling of a Solar Air Collector Heat Transfer Coefficient with Regression Algorithms. *Firat Univ. J. Exp. Comput. Eng.* **2022**, *1*, 14–23. [CrossRef]
37. Atila, O.; Şengür, A. Attention Guided 3D CNN-LSTM Model for Accurate Speech Based Emotion Recognition. *Appl. Acoust.* **2021**, *182*, 108260. [CrossRef]

38. Niu, Z.; Zhong, G.; Yu, H. A Review on the Attention Mechanism of Deep Learning. *Neurocomputing* **2021**, *452*, 48–62. [CrossRef]
39. Abdelaziz Ismael, S.A.; Mohammed, A.; Hefny, H. An Enhanced Deep Learning Approach for Brain Cancer MRI Images Classification Using Residual Networks. *Artif. Intell. Med.* **2020**, *102*, 101779. [CrossRef]
40. Zhao, Y.; Liu, Y.; Huang, W. Prediction Model of HBV Reactivation in Primary Liver Cancer—Based on NCA Feature Selection and SVM Classifier with Bayesian and Grid Optimization. In *Proceedings of the 2018 IEEE 3rd International Conference on Cloud Computing and Big Data Analysis (ICCCBDA), Chengdu, China, 20–22 April 2018*; IEEE: New York, NY, USA, 2018; pp. 547–551.
41. Muqeet, H.A.; Liaqat, R.; Hussain, A.; Sajjad, I.A.; Ahmad, H.; Mehmood, A. Regularized NCA Based Prominent Feature Selection for Load Identification Using Boosted Tree Classifier. 2024. Available online: https://assets-eu.researchsquare.com/files/rs-4151725/v1_covered_d7165e06-8243-4a13-81e3-1fe6a5ad9d35.pdf (accessed on 1 March 2025).
42. Khatri, S.; Bansal, P. Hyperparameter Tuning and Validation of Neural Network Model for Software Effort Estimation with NCA Based Feature Selection. In *Proceedings of the 2024 IEEE International Conference on Information Technology, Electronics and Intelligent Communication Systems (ICITEICS), Bangalore, India, 28–29 June 2024*; IEEE: New York, NY, USA, 2024; pp. 1–6.
43. Demir, F.; Siddique, K.; Alswaitti, M.; Demir, K.; Sengur, A. A Simple and Effective Approach Based on a Multi-Level Feature Selection for Automated Parkinson's Disease Detection. *J. Pers. Med.* **2022**, *12*, 55. [CrossRef]
44. Yildiz, A.M.; Gun, M.V.; Yildirim, K.; Keles, T.; Dogan, S.; Tuncer, T.; Acharya, U.R. SKLBP14: A New Textural Environmental Sound Classification Model Based on a Squarekernelled Local Binary Pattern. *Firat Univ. J. Exp. Comput. Eng.* **2023**, *2*, 46–54. [CrossRef]
45. Baygin, M.; Yaman, O.; Tuncer, T.; Dogan, S.; Barua, P.D.; Acharya, U.R. Automated Accurate Schizophrenia Detection System Using Collatz Pattern Technique with EEG Signals. *Biomed. Signal Process. Control* **2021**, *70*, 102936. [CrossRef]
46. Tuncer, T.; Dogan, S.; Subasi, A. EEG-Based Driving Fatigue Detection Using Multilevel Feature Extraction and Iterative Hybrid Feature Selection. *Biomed. Signal Process. Control* **2021**, *68*, 102591. [CrossRef]
47. Turkoglu, M. COVIDetectioNet: COVID-19 Diagnosis System Based on X-Ray Images Using Features Selected from Pre-Learned Deep Features Ensemble. *Appl. Intell.* **2021**, *51*, 1213–1226. [CrossRef]
48. Tuncer, T.; Ertam, F. Neighborhood Component Analysis and ReliefF Based Survival Recognition Methods for Hepatocellular Carcinoma. *Phys. A Stat. Mech. Its Appl.* **2020**, *540*, 123143. [CrossRef]
49. Demir, F.; Taşcı, B. An Effective and Robust Approach Based on R-CNN + LSTM Model and NCAR Feature Selection for Ophthalmological Disease Detection from Fundus Images. *J. Pers. Med.* **2021**, *11*, 1276. [CrossRef]
50. Fadli, V.F.; Herlistiono, I.O. Steel Surface Defect Detection Using Deep Learning. *Int. J. Innov. Sci. Res. Technol.* **2020**, *5*, 244–250. [CrossRef]
51. Guo, X.; Liu, X.; Królczyk, G.; Sulowicz, M.; Glowacz, A.; Gardoni, P.; Li, Z. Damage Detection for Conveyor Belt Surface Based on Conditional Cycle Generative Adversarial Network. *Sensors* **2022**, *22*, 3485. [CrossRef]
52. Chigateri, K.B.; Hebbale, A.M. A Steel Surface Defect Detection Model Using Machine Learning. *Mater. Today Proc.* **2023**, *100*, 51–58. [CrossRef]
53. Hu, J.; Shen, L.; Sun, G. Squeeze-and-Excitation Networks. In Proceedings of the IEEE Conference on Computer Vision and Pattern Recognition, Salt Lake City, UT, USA, 18–23 June 2018; pp. 7132–7141.
54. Yeung, C.C.; Lam, K.M. Efficient Fused-Attention Model for Steel Surface Defect Detection. *IEEE Trans. Instrum. Meas.* **2022**, *71*, 2510011. [CrossRef]
55. Tian, S.; Huang, P.; Ma, H.; Wang, J.; Zhou, X.; Zhang, S.; Zhou, J.; Huang, R.; Li, Y. CASDD: Automatic Surface Defect Detection Using a Complementary Adversarial Network. *IEEE Sens. J.* **2022**, *22*, 19583–19595. [CrossRef]
56. Yi, L.; Li, G.; Jiang, M. An End-to-End Steel Strip Surface Defects Recognition System Based on Convolutional Neural Networks. *Steel Res. Int.* **2017**, *88*, 176–187. [CrossRef]

applied sciences

MDPI

Article

ETAFHrNet: A Transformer-Based Multi-Scale Network for Asymmetric Pavement Crack Segmentation

Chao Tan, Jiaqi Liu, Zhedong Zhao, Rufei Liu *, Peng Tan, Aishu Yao, Shoudao Pan and Jingyi Dong

College of Geodesy and Geomatics, Shandong University of Science and Technology, Qingdao 266590, China; 202211030326@sdust.edu.cn (C.T.)
* Correspondence: liurufei@sdust.edu.cn

Abstract: Accurate segmentation of pavement cracks from high-resolution remote sensing imagery plays a crucial role in automated road condition assessment and infrastructure maintenance. However, crack structures often exhibit asymmetry, irregular morphology, and multi-scale variations, posing significant challenges to conventional CNN-based methods in real-world environments. Specifically, the proposed ETAFHrNet focuses on two predominant pavement-distress morphologies—linear cracks (transverse and longitudinal) and alligator cracks—and has been empirically validated on their intersections and branching patterns over both asphalt and concrete road surfaces. In this work, we present ETAFHrNet, a novel attention-guided segmentation network designed to address the limitations of traditional architectures in detecting fine-grained and asymmetric patterns. ETAFHrNet integrates Transformer-based global attention and multi-scale hybrid feature fusion, enhancing both contextual perception and detail sensitivity. The network introduces two key modules: the Efficient Hybrid Attention Transformer (EHAT), which captures long-range dependencies, and the Cross-Scale Hybrid Attention Module (CSHAM), which adaptively fuses features across spatial resolutions. To support model training and benchmarking, we also propose QD-Crack, a high-resolution, pixel-level annotated dataset collected from real-world road inspection scenarios. Experimental results show that ETAFHrNet significantly outperforms existing methods—including U-Net, DeepLabv3+, and HRNet—in both segmentation accuracy and generalization ability. These findings demonstrate the effectiveness of interpretable, multi-scale attention architectures in complex object detection and image classification tasks, making our approach relevant for broader applications, such as autonomous driving, remote sensing, and smart infrastructure systems.

Keywords: pavement crack segmentation; transformer neural networks; multi-scale feature fusion; global attention mechanism; high-resolution remote sensing; deep learning; infrastructure monitoring; interpretable classification

1. Introduction

Pavement cracks are critical indicators of road infrastructure integrity, and their early and accurate detection plays a vital role in supporting preventive maintenance, extending service life, and ensuring traffic safety. According to global statistics, surface cracks contribute to over 30% of road-related traffic accidents annually [1]. If left unrepaired, they allow moisture penetration, accelerating substructure deterioration and significantly increasing maintenance costs. Studies have shown that untreated cracks can raise annual road maintenance expenditures by approximately 15%.

Conventional manual inspections suffer from low efficiency and high subjectivity. Reported detection rates fall below 80%, with false detection rates exceeding 30% [2]. While experienced inspectors can recognize visible damage, manual approaches are difficult to scale and insufficiently accurate for large road networks. In contrast, automated vision-based systems have reduced false positive rates to under 5% [3], offering a promising direction for smart pavement monitoring.

Traditional methods based on threshold segmentation [4] or edge detection [5] perform poorly under complex lighting and noise conditions. The emergence of deep learning has led to substantial progress in crack segmentation. Encoder-decoder networks such as U-Net [6] enable end-to-end detection. DeepCrack [7], for instance, achieves high IoU through multi-scale fusion but struggles to retain fine-grained structural details.

To improve spatial resolution, HRNet [8] was introduced, maintaining high-resolution representations via parallel branches. Yang et al. [9] and Fan et al. [10] adopted multi-resolution and adaptive thresholding strategies, yet their models still underperform in detecting fine or net-like cracks.

In recent years, researchers have increasingly adopted attention mechanisms and Transformer-based architectures to improve global perception and feature representation in crack segmentation tasks. Chen et al. [11] and Wang et al. [12] introduced channel and non-local spatial attention, significantly enhancing discriminability and context awareness. However, these methods typically incur high computational overhead.

Further developments such as SENet [13], CBAM [14], and Pyramid Attention Networks [15] improved adaptability to crack morphology but still lack flexible weighting mechanisms and robust generalization under noisy backgrounds.

In the Transformer domain, ViT [16] enables long-range modeling but demands extensive computation. Swin Transformer [17] reduces complexity via window partitioning but compromises spatial continuity. SegFormer [18] merges CNN and Transformer strengths, achieving balanced performance, but fixed attention fusion often weakens fine-detail segmentation [19,20].

Recent improvements by Zheng et al. [21], Ding et al. [22], and Huang et al. [23] demonstrate progress in contextual interaction and structural awareness. However, many models still struggle to address high-resolution, multi-scale, and complex crack geometries encountered in practical deployments.

Two major challenges remain unresolved:

(1) **Accurate identification of intersecting cracks.** In real scenarios, cracks often branch or intersect. Without sufficient receptive field or contextual awareness, models tend to miss or misclassify these areas [24].
(2) **Continuous modeling of long-range cracks.** Cracks are typically thin and extended. In the absence of strong global context modeling, segmentation results become fragmented, particularly under high-resolution or multi-scale settings [7].

To address these issues, we propose a novel segmentation framework—ETAFHrNet (Efficient Transformer-Enhanced and Adaptive Fusion Attention Network)—which integrates convolutional and Transformer paradigms to balance accuracy and efficiency.

(1) **Global-local collaborative feature modeling:** We introduce an Efficient Hybrid Attention Transformer (EHAT) module into HRNet's high-resolution branches, combining axial positional encoding and window attention to capture long-range dependencies while controlling computation [25,26].
(2) **Adaptive multi-scale fusion:** A novel Cross-Scale Hybrid Attention Module (CSHAM) adaptively weights spatial and directional features through cascaded axial and cross-scale attention, enhancing the detection of intersecting or subtle crack patterns [27,28].

The overall workflow of the proposed method is illustrated in Figure 1, which outlines the entire pipeline from data acquisition to output segmentation. This structured design ensures reproducibility and operational scalability in pavement crack detection tasks.

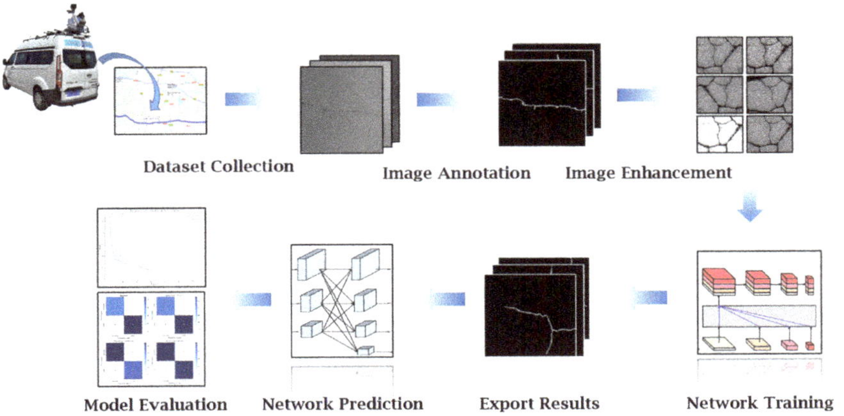

Figure 1. Workflow of the proposed pavement crack detection framework.

2. Related Work

In recent years, deep learning has demonstrated substantial potential in the domain of image segmentation, particularly within structural health monitoring applications. Among emerging trends, convolutional neural networks (CNNs) enhanced by attention mechanisms and Transformer-based architectures have attracted increasing research attention. Against this backdrop, this section presents a structured review of recent advancements in pavement crack detection, organized from three key perspectives: (1) the evolution of classical segmentation models; (2) the development and refinement of attention mechanisms; (3) recent breakthroughs in global context modeling. In addition, we critically assess the applicability and limitations of these methods in addressing the unique challenges posed by crack detection, including scale variation, spatial discontinuity, and background complexity.

2.1. Segmentation Model Evolution

Regarding the evolution of classical segmentation models, Fully Convolutional Networks (FCNs) [29] were the first to introduce end-to-end, pixel-level prediction frameworks, thereby laying the groundwork for modern semantic segmentation. U-Net [30] advanced this concept by proposing an encoder-decoder architecture with skip connections, achieving notable success in biomedical image segmentation and inspiring subsequent model designs. Building on these foundations, multi-scale feature fusion strategies have been widely adopted to further enhance segmentation performance. The DeepLab series [11] employed atrous (dilated) convolutions to expand the receptive field without compromising spatial resolution. PSPNet [31] introduced a Pyramid Pooling Module to effectively capture multi-scale contextual information. More recently, HRNet improved segmentation accuracy by maintaining high-resolution representations through parallel multi-branch architectures, enabling the precise localization of fine structural details.

Zhang et al. [32] applied HRNet to pavement crack detection and demonstrated the advantages of multi-resolution feature fusion for identifying elongated cracks under complex background conditions. However, conventional CNN-based models often rely on local convolutional operations, which struggle to simultaneously achieve global semantic understanding and fine-grained representation, particularly when cracks are morphologically diverse, sparsely distributed, or embedded in noisy surfaces [10]. For example, although U-Net preserves low-level features through skip connections, its fixed receptive field restricts its ability to capture the global topological continuity of cracks across varying scales. DeepLabv3+ extends contextual awareness through atrous convolutions, yet remains vulnerable to false positives caused by background surface noise and exhibits inadequate continuity modeling for slender, elongated cracks. PSPNet incorporates pyramid pooling for multi-scale context aggregation, but its coarse-grained feature integration tends to overlook small or subtle crack patterns. Even though HRNet excels at maintaining high-resolution features via parallel multi-branch structures, its reliance on traditional convolutions limits its capacity to model long-range dependencies in complex scenes.

Recent work by Yin et al. [33] introduced DCRNet, a dual-context residual network that jointly models local detail and global structure using parallel pathways. This dual-path design aligns closely with our use of the EHAT and CSHAM modules, which aim to enhance crack connectivity and multiscale representation. DCRNet has shown strong performance in capturing complex crack morphologies and thus provides a meaningful comparative reference for dual-context segmentation architectures.

These observations underscore two persistent challenges in CNN-based crack detection: (1) insufficient global perception to capture long-range crack structures, and (2) limited local feature representation for accurately identifying fine, fragmented, or intersecting cracks.

2.2. Attention Mechanisms

The introduction of attention mechanisms has provided new opportunities for addressing the challenges of feature selection and fusion in pavement crack detection. Early methods such as SENet [13] utilized global average pooling to capture inter-channel dependencies and dynamically reweight channel responses. However, due to the absence of spatial interaction, SENet remains insufficient for detecting elongated or spatially distributed crack structures. CBAM [14], which incorporates both channel and spatial attention, improves segmentation performance by focusing on locally salient features. Yet, it still lacks the capability to model long-range spatial dependencies, which are critical for capturing the continuity of dispersed crack segments.

To enhance global context modeling, DANet [20] introduced a dual-path attention structure, while non-local modules [34] employed self-attention mechanisms to establish pixel-level global correlations. Despite these advancements, many existing approaches rely on fixed-weight fusion strategies. For instance, in the work by Li et al. [35], CBAM and the non-local module are combined in series, yet the static integration scheme fails to adapt to the morphological diversity of cracks. In contrast, Guo et al. [36] proposed a dynamic convolutional attention network that employs learnable weights to adaptively assign attention across features, offering a promising approach for handling multi-scale and complex crack patterns.

Li et al. [37] proposed CrackCLF, a closed-loop feedback-based segmentation network that iteratively refines predictions by incorporating previous outputs as inputs. This dynamic correction mechanism complements our adaptive attention design and represents a promising direction for improving segmentation stability in noisy environments.

The design of attention mechanisms is especially critical in pavement crack detection, where the structures of interest are typically slender, elongated, and oriented in diverse directions. Detection algorithms must therefore balance the preservation of local detail with the need for global continuity [38]. While traditional attention modules such as CBAM are effective in enhancing local contrast, they often fail to establish relationships between spatially separated crack fragments. Recent studies suggest that directional and topological cues are key to improving detection accuracy. In particular, axial attention has been shown to enhance the recognition of horizontal and vertical crack components by independently modeling one-dimensional spatial dependencies [39].

Furthermore, Chen et al. [40] proposed an edge-aware attention network that explicitly enhances boundary preservation and continuity through guided refinement. This is particularly relevant to the directional and topological modeling objectives of our EHAT and CSHAM modules.

Nonetheless, many attention mechanisms continue to employ fixed-weight configurations, which limits their adaptability across diverse scenes. In environments where reticular and linear cracks coexist, this rigidity results in suboptimal segmentation performance [22]. Therefore, the development of dynamically adaptive attention mechanisms capable of modeling multi-scale, morphologically diverse crack structures remains an open and significant research challenge.

2.3. Transformer Architectures

In recent years, the remarkable performance of the Vision Transformer (ViT) in computer vision has spurred widespread exploration of global context modeling methods. ViT achieves holistic semantic representation by segmenting images into fixed-size patches and applying a multi-head self-attention mechanism to process them. However, its substantial computational cost and reliance on large-scale datasets limit its practicality in real-world deployment scenarios.

To balance accuracy and efficiency, a variety of Transformer-CNN hybrid architectures have been proposed. For instance, TransUNet [41] embeds local features extracted by convolutional layers into a Transformer encoder while employing skip connections to preserve spatial detail. CMT [42] introduces a dual-branch structure that facilitates dynamic interactions between local and global representations. Similarly, Mobile-Former [43] adopts a lightweight architecture to reduce computational overhead for mobile and embedded scenarios.

While these hybrid approaches have demonstrated success in general semantic segmentation tasks, their applicability to pavement crack detection remains limited. For example, the window-based partitioning strategy in the Swin Transformer [17] can disrupt the continuity of linear crack patterns, impairing segmentation accuracy. Likewise, the dual-branch structure in Mobile-Former incurs significant memory consumption when applied to high-resolution inputs [44]. Moreover, many hybrid models are based on the U-Net framework, which may not align well with the architectural design of HRNet, particularly in terms of maintaining high-resolution feature representations throughout the network [8].

As a representative Transformer-CNN fusion method, SegFormer has achieved a mean Intersection over Union (mIoU) of 79.5% in general segmentation tasks, attributed to its robust global context modeling capabilities [18]. However, its window partitioning mechanism may introduce discontinuities in crack representation, particularly under ultra-high-resolution inputs. Additionally, its ability to support real-time detection remains limited. In response, several lightweight Transformer modules have been proposed to reduce computational burden through local window attention and dimensionality reduction strategies.

Despite these improvements, a fundamental challenge persists: how to effectively represent directional crack features while maintaining global perceptual awareness [45]. Recent advances in axial positional encoding and directional feature enhancement mechanisms offer promising solutions, particularly for capturing the elongated and linear nature of pavement cracks [36]. Future research should continue to explore adaptive attention mechanisms and high-resolution feature preservation strategies. In particular, integrating the local sensitivity of CNNs with the global dependency modeling strengths of Transformers represents a promising direction for achieving both fine-grained precision and real-time performance in practical pavement crack detection systems.

3. Methods

3.1. ETAFHrNet Architecture

This paper presents a novel network architecture, termed the Efficient Transformer-Enhanced and Adaptive Fusion High-Resolution Network (ETAFHrNet). The overall architecture is illustrated in Figure 2. Building upon the HRNet framework, the proposed model preserves HRNet's strength in maintaining multi-resolution feature representations, while integrating two key innovations: the Efficient Hybrid Attention Transformer (EHAT) and the Cross-Scale Hybrid Attention Module (CSHAM). These components are specifically designed to enhance the network's capacity for crack representation by improving global context modeling and multi-scale feature fusion.

The architecture comprises three primary components. First, the HRNet backbone extracts multi-resolution features through parallel branches, maintaining high-resolution feature flow while generating rich semantic representations. Second, the EHAT module is embedded into the high-resolution branch to perform lightweight long-range dependency modeling. This is achieved through a combination of adaptive channel dimensionality reduction, axial positional encoding, and local window attention-mechanisms that are particularly effective in enhancing the perception of linear and directional crack structures. Third, following feature alignment via cross-resolution upsampling, the CSHAM module conducts cross-scale adaptive fusion and directional enhancement of multi-level features. This ensures that the segmentation head receives a comprehensive, high-resolution feature representation, enabling the generation of accurate prediction maps aligned with the input resolution.

To overcome the limitations of conventional HRNet in pavement crack detection—specifically, its limited global semantic modeling and rigid feature fusion—this study introduces two architectural innovations. First, the EHAT module improves linear feature representation by integrating axial positional encoding and local window attention. Its hybrid MLP structure combines the local inductive bias of convolutional operations with the global modeling capacity of Transformers, making it highly effective in capturing crack features across multiple orientations and scales, particularly in complex or subtle crack scenarios. Second, the CSHAM module applies a cross-scale attention mechanism to adaptively weight multi-resolution features and leverages axial attention to enhance directional feature expression. This design alleviates the information loss often caused by static fusion in conventional HRNet and performs robustly in scenes where slender, intersecting, and multi-scale cracks coexist.The following subsections provide a detailed explanation of the proposed EHAT and CSHAM modules.

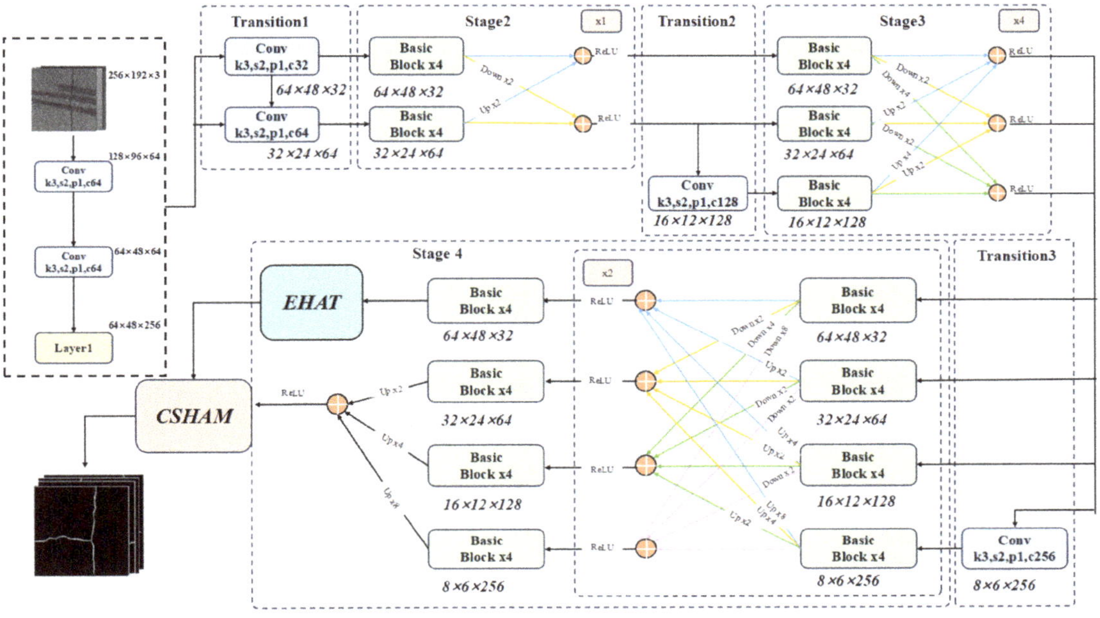

Figure 2. The ETAFHrNet architecture.

3.2. Efficient Hybrid Attention Transformer (EHAT) Module

This module takes an input feature map of shape $B \times C \times H \times W$, where B denotes the batch size, C, the number of channels, and H and W, the height and width, respectively. To preserve the linear structural features of cracks while reducing computational complexity, the Efficient Hybrid Attention Transformer (EHAT) employs a series of optimization strategies, including adaptive channel reduction [46], axial enhancement, and local window attention [47]. An overview of the EHAT module's architecture is presented in Figure 3.

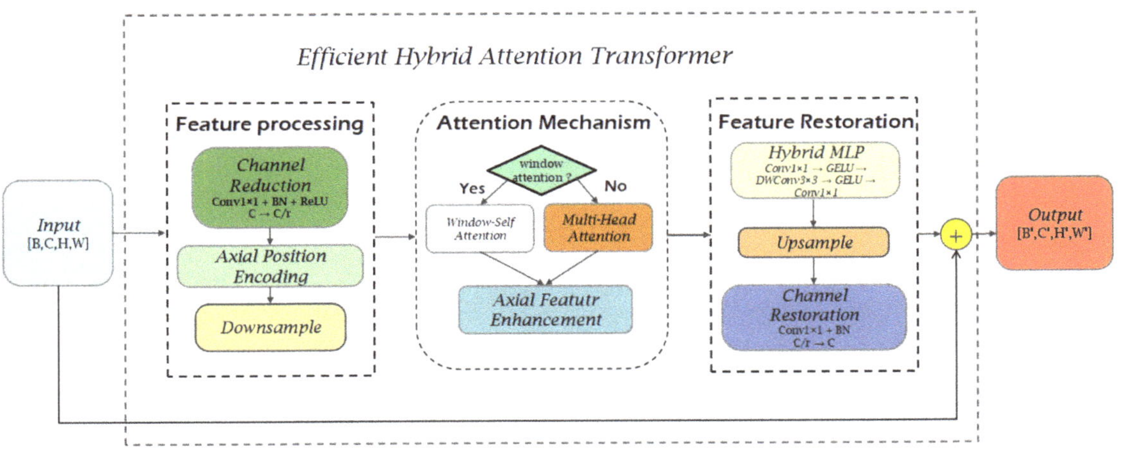

Figure 3. The Efficient Hybrid Attention Transformer (EHAT) Module.

A 1×1 convolution reduces the input channels from C to $C' = (C_r - c)/r$, where r is the reduction ratio, C_r the base number of reduced channels, and c a learnable channel bias. Spatial downsampling via bilinear interpolation, with a ratio of s, yields dimensions $H' = H/s$ and $W' = W/s$, thereby compressing both spatial and channel dimensions to alleviate the computational burden. Axial positional encoding is subsequently applied to independently enhance features along the horizontal and vertical axes, introducing directional priors well suited for representing elongated crack structures.

$$F_{\text{axial}} = \text{Concat}(F_h + P_h, F_v + P_v) \tag{1}$$

The downsampled feature map is partitioned into two components, F_h and F_v, corresponding to horizontal and vertical orientations. Learnable positional parameters P_h and P_v are added to each, after which, the results are concatenated via `Concat()`. This axial positional encoding introduces explicit directional cues, enhancing the model's capacity to identify and preserve linear crack features. To balance computational efficiency with structural sensitivity, EHAT employs local attention within each axis, enabling efficient modeling of elongated patterns without incurring the overhead of full self-attention.

$$\text{Attention}(Q, K, V) = \text{Softmax}\left(\frac{QK^T}{\sqrt{d_k}}\right)V \tag{2}$$

Here, Q, K, and V denote the query, key, and value matrices, respectively, and d_k represents the dimensionality of the query vectors. The local window attention mechanism divides the feature map into non-overlapping regions, where self-attention is calculated independently within each window. This formulation preserves local structural integrity while significantly reducing the computational cost associated with full self-attention. To further enhance directional sensitivity, EHAT integrates an axial feature enhancement module, refining the representation of elongated crack patterns across horizontal and vertical orientations.

$$F_{\text{enhanced}} = F_{\text{axial}} + \text{Conv}_{\text{axial}}(F_{\text{axial}}) \tag{3}$$

Here, $\text{Conv}_{\text{axial}}(\cdot)$ denotes a convolution operation applied along the axial direction, designed to further refine crack-related feature representations. Departing from the standard Transformer paradigm, EHAT adopts a hybrid MLP structure [48], integrating convolutional layers with multilayer perceptrons [49]. This design leverages the local inductive bias of convolutions alongside the global modeling capacity of MLPs, enhancing the network's ability to capture both fine-grained details and long-range dependencies.

$$F_{\text{hybrid}} = \text{MLP}(F_{\text{enhanced}}) + \text{Conv}(F_{\text{enhanced}}) \tag{4}$$

$\text{MLP}(\cdot)$ denotes a multilayer perceptron (feedforward network), and $\text{Conv}(\cdot)$ denotes a conventional convolution operation. By summing the outputs of both operations, the model effectively fuses global context with fine-grained spatial features.

The axial feature enhancement module reinforces the linear characteristics of cracks along both horizontal and vertical directions. Meanwhile, the hybrid MLP architecture combines the inductive biases of convolution with the expressive capacity of Transformers. To complete the EHAT module's processing pipeline, feature maps are first upsampled to their original resolution, followed by a 1×1 convolution to align channel dimensions.

The EHAT module incorporates several technical innovations within its overall architecture: it reduces computational complexity through adaptive channel reduction and local window attention; enhances directional feature perception via axial positional encoding and feature enhancement mechanisms; and integrates the local inductive bias of convolution with the global modeling capacity of Transformers through a hybrid MLP structure. These design innovations enable the EHAT module to maintain a lightweight structure while substantially enhancing the network's capacity to detect cracks across diverse orientations and scales. It performs particularly well in complex backgrounds and subtle crack scenarios, providing a robust feature representation foundation for high-precision pavement crack segmentation.

3.3. Cross-Scale Hybrid Attention Module (CSHAM)

In the original HRNet architecture, the multi-scale feature fusion stage performs feature alignment across different resolution branches via upsampling, followed by direct fusion through summation or concatenation. This rigid fusion strategy lacks both adaptive weighting across scales and directional feature enhancement, which limits its effectiveness in pavement crack segmentation, particularly for elongated structures and scenes involving the coexistence of multi-scale cracks. In such scenarios, critical morphological information is often lost due to the uniform treatment of features with varying semantic granularity.

To overcome these limitations, we introduce the Cross-Scale Hybrid Attention Module (CSHAM) into the multi-scale fusion stage of HRNet. Specifically, after all feature maps are upsampled to a unified spatial resolution, CSHAM is inserted in place of naive fusion (as illustrated in Figure 4), enabling both cross-scale adaptive fusion [50] and axial attention enhancement [39]. This design ensures that the segmentation head receives a comprehensive, structurally-aware representation that integrates multi-scale contextual information while preserving directional cues critical for detecting complex crack morphologies.

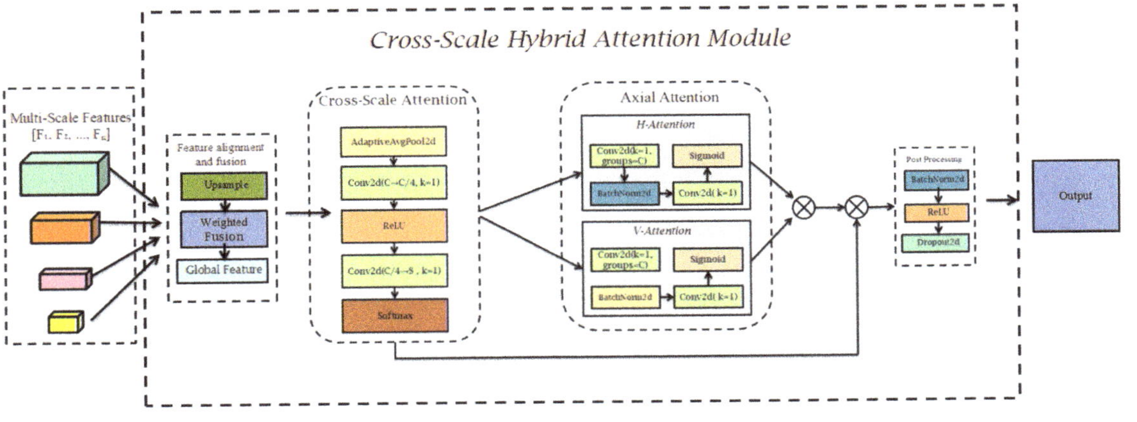

Figure 4. The Cross-Scale Hybrid Attention Module (CSHAM).

The Cross-Scale Hybrid Attention Module (CSHAM) adopts a hierarchical connection structure designed to perform adaptive scale-wise feature weighting alongside directional feature enhancement. The process begins with a cross-scale attention mechanism, which evaluates the relative importance of each scale-specific feature by aggregating global contextual information. The computation is formally expressed as:

$$G = \frac{1}{N}\sum_{i=1}^{N} F_i \tag{5}$$

Let F_i denote the feature map at scale i, where $i = 1, 2, \ldots, N$ and N is the total number of scales. To compute the importance of each scale, a channel dimension reduction operation is first applied, followed by a weight prediction module that assigns learnable importance scores. These weights are then used to reweight the corresponding feature maps, enabling the network to dynamically emphasize informative scales while suppressing less relevant representations.

$$W = \text{Softmax}(W_2\,\delta(W_1\,G)) \tag{6}$$

$$F_{\text{fused}} = \sum_{i=1}^{N} W_i \cdot F_i \tag{7}$$

Here, W_1 and W_2 are learnable projection matrices used for dimensionality reduction and expansion, respectively, and δ represents the ReLU activation function. Once the attention weights are computed and scale-wise feature reweighting is performed, a fused feature map F_{fused} is obtained. To further enhance directional awareness, directional attention is applied by employing one-dimensional convolutions to generate attention maps along the horizontal and vertical axes. These attention maps are then element-wise multiplied with F_{fused} to selectively amplify features aligned with directional crack patterns. This process is formally defined as:

$$F_{\text{axial}} = \sigma(\text{Conv}_h(F_{\text{fused}})) \odot \sigma(\text{Conv}_v(F_{\text{fused}})) \tag{8}$$

In this context, Conv_h and Conv_v refer to one-dimensional convolution operations performed along the horizontal and vertical axes, respectively. The function σ represents the Sigmoid activation, and $\odot$ denotes element-wise multiplication. This design is particularly effective in enhancing the model's sensitivity to linear crack structures, regardless of orientation.

During backpropagation, the gradients of the cross-scale attention weights can be computed as:

$$\frac{\partial L}{\partial W_i} = \frac{\partial L}{\partial F_{\text{out}}} \cdot \frac{\partial F_{\text{out}}}{\partial F_{\text{fused}}} \cdot F_i \tag{9}$$

Let L denote the loss function, which quantifies the discrepancy between model predictions and ground truth labels. The parameter W_i represents the learnable attention weight for scale i, while F_i denotes the corresponding input feature map. Through gradient backpropagation, the model dynamically adjusts W_i to optimize the importance of each scale relative to the segmentation objective. The fused feature map F_{fused} encapsulates aggregated multi-scale representations, and the final output produced by the CSHAM module is represented as F_{out}. This formulation enables the network to selectively enhance both scale-sensitive and directionally discriminative features, thereby improving segmentation accuracy and promoting continuity in predicted crack structures.

The CSHAM module serves as a key component in facilitating multi-scale feature fusion within the HRNet architecture. Beyond optimizing the adaptive integration of features across resolutions, it significantly strengthens directional feature representation. By capturing diverse crack morphologies and preserving fine structural details, CSHAM contributes directly to improved segmentation performance and more precise edge localization across a wide range of road surface conditions.

4. Experimental Details

4.1. Dataset Preparation

As a pixel-level classification task, the performance of image segmentation models is highly dependent on the quality and diversity of the training dataset. However, existing publicly available road crack detection datasets often lack finely annotated templates capable of capturing the wide morphological variability of cracks across diverse real-world conditions. To address this gap, we constructed a dedicated multi-scene segmentation dataset focused on road surface cracks, referred to as the QD-Crack dataset. This dataset is based on high-resolution road surface imagery collected by professional pavement inspection vehicles operating on expressways in Shandong Province, China, since May 2023. All data collection activities were conducted with the authorization of relevant municipal authorities. To ensure data privacy and regulatory compliance, all original images were preprocessed to remove identifiable elements, such as licence plates and prominent landmarks. The base dataset comprises 500 high-resolution images, captured under a wide range of environmental conditions, including varying lighting, pavement materials, and crack types. These images reflect diverse forms of pavement distress and are stored in JPG format. Annotation was performed by a team of experienced road maintenance engineers—each with over three years of professional experience—using the Labelme tool for detailed, vector-based labeling of crack morphology. To ensure annotation quality and consistency, all labels were cross-verified by two independent inspection engineers. Discrepancies were resolved through panel-based expert review. The overall dataset construction workflow is illustrated in Figure 5. The QD-Crack dataset was collected by the authors from municipal roads in Qingdao, China. While it is not currently publicly available, it can be accessed upon reasonable request to the corresponding author for research purposes.

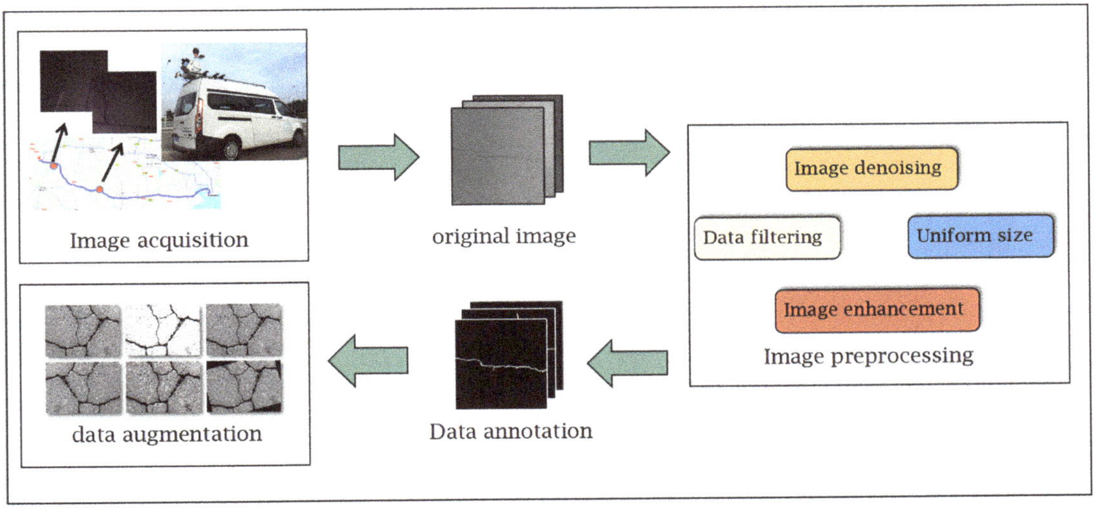

Figure 5. Dataset construction process.

Given the high cost and labour intensity of pixel-level annotation, we adopted a semi-automatic labeling strategy inspired by the approach of Jia et al. [51]. This method focuses on annotating the primary crack structures rather than the intact road surface, thereby improving labeling efficiency while preserving semantic relevance. Initial annotations were generated with the assistance of edge detection algorithms, which provided a contour-based approximation of crack boundaries. The final annotated dataset comprises a JSON file containing approximately 13.5 million labeled points, subsequently converted into PNG-format semantic segmentation masks using a custom-developed Python 3.10 script. To examine the influence of annotation granularity on model performance, we designed two distinct labeling schemes: (1) a binary scheme with two categories: background and crack; (2) a three-class scheme, comprising background, linear cracks, and alligator (reticular) cracks. Representative examples from labeling schemes are shown in Figure 6.

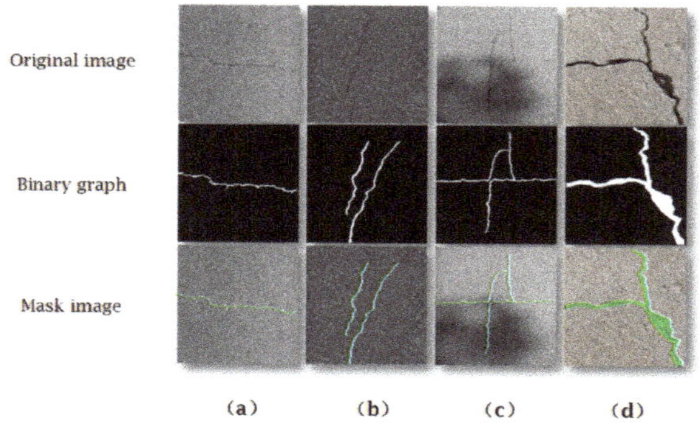

Figure 6. Examples of labeling schemes for different crack types. Each column from (**a–d**) represents a complete sample group, consisting of the following: top row—original pavement image, middle row—binary annotation, bottom row—overlay mask. Specifically, (**a**) transverse crack, (**b**) longitudinal crack, (**c**) alligator crack on asphalt pavement, and (**d**) intersecting crack on cement pavement.

To enhance dataset utility and improve the robustness and generalization capability of the trained models, we applied a set of essential image preprocessing procedures, including image enhancement and geometric correction. A comprehensive data augmentation pipeline was implemented, incorporating random angle rotation, brightness adjustment, contrast enhancement, sharpness optimization, and horizontal flipping. As a result, the dataset was expanded from 500 to a total of 2500 samples. These augmentation techniques not only increase data diversity but also emulate complex real-world engineering conditions. For instance, random rotation (within ±15°) allows the model to recognise cracks from multiple viewing angles; brightness and contrast adjustments enhance texture visibility under variable lighting; and sharpness optimization amplifies edge contrast, thereby improving the distinction between crack regions and the background. An illustration of these effects is provided in Figure 7. Additionally, horizontal flipping augments data volume while mitigating directional bias, encouraging the model to generalize across diverse crack orientations and morphologies.

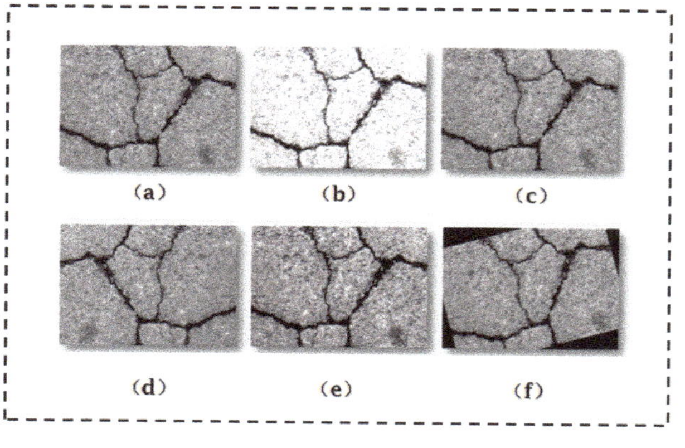

Figure 7. Image preprocessing visualization. The images from top-left to bottom-right are as follows: (**a**) original image, (**b**) contrast-enhanced, (**c**) brightness-adjusted, (**d**) horizontally flipped, (**e**) sharpness-optimized, and (**f**) randomly rotated.

4.2. Training Parameters and Methods

To ensure the accuracy and reproducibility of the experiments, the detailed configuration of the experimental environment is summarized in Table 1.

Table 1. Model training configuration.

Configuration Items	Configuration
Operating System	Windows 11
Deep Learning Framework	PyTorch 1.10.0
Processor	Intel Core i7-12700k
RAM	32 GB
GPU	NVIDIA GeForce RTX 3070 Ti (8 GB)
GPU Memory	8 GB
CUDA Version	11.3

During the experimental procedure, the dataset was divided into a training set and a test set at a fixed ratio of 8:2. Model parameters were iteratively optimized using the training set, whereas the test set was reserved for assessing generalization capability. To efficiently manage GPU memory limitations, the batch size was configured to 8, enabling effective utilization of the available computational resources. Based on prior experimental experience, the number of training epochs was uniformly set to 120, as the loss function consistently converged near this point across multiple configurations. The convergence behavior is visualized in Figure 8, which illustrates the decline and stabilization of the loss function across epochs. The model achieves a stable convergence state by approximately the 120th epoch, validating the effectiveness of the selected training schedule.

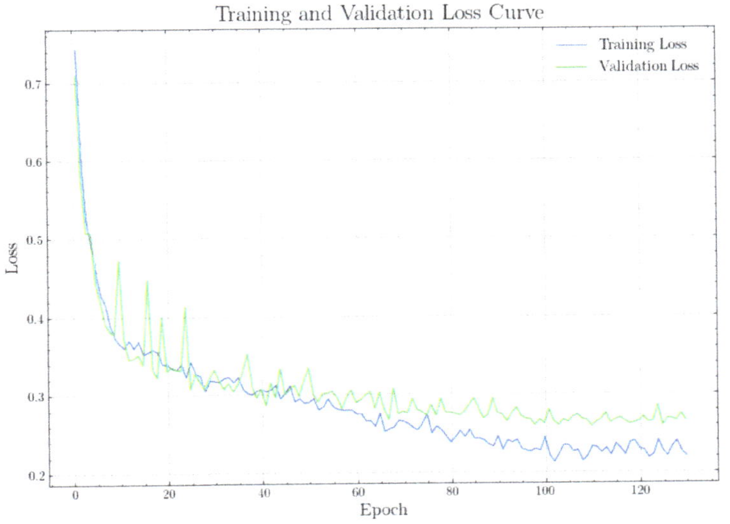

Figure 8. The trend of loss function with the number of training rounds.

In image segmentation tasks, accurate delineation of object contours often relies on spatially consistent feature distributions. Building upon this observation, we incorporated a transfer learning strategy to enhance model learning efficiency. The core principle of transfer learning lies in reusing knowledge—specifically, pre-trained model weights—from one task to accelerate learning in a related but distinct target task, much like how humans transfer prior experience to new problems. In this work, the model parameters were initially pre-trained on a large-scale crack detection dataset, and subsequently fine-tuned on the target dataset to adapt to the specific task requirements.

The SDNET2018 dataset [52] was selected as the source for pre-training. Comprising over 56,000 annotated concrete crack images, SDNET2018 is widely recognized in the field for its utility in training, validation, and benchmarking of crack detection algorithms. Experimental results demonstrate that this transfer learning approach [53] not only accelerates convergence but also yields significant improvements in segmentation accuracy. Owing to the visual feature similarities shared across diverse real-world objects, this strategy closely aligns with human perceptual learning processes.

To fully exploit the benefits of pre-trained knowledge, we adopted a freeze-thaw training strategy [54], as opposed to random weight initialization. Given that the network backbone is responsible for extracting generalizable low-level features, its parameters were initially frozen, while the remaining layers were fine-tuned on the target data. During the mid-to-late training phases, the backbone was gradually unfrozen to allow full network optimization and better task adaptation.

The initial learning rate was set to 0.0001 and dynamically adjusted using a cosine annealing schedule [55] to improve convergence stability and efficiency. To further stabilize training, we set the momentum parameter to 0.975 and employed the Adam optimizer [56], which adaptively adjusts learning rates by incorporating both first- and second-order moment estimates of the gradients.

4.3. Methods for Evaluation

Evaluations were carried out on the QD-Crack dataset as well as other publicly available crack segmentation datasets to comprehensively assess both the performance gains and generalization ability of our model. To assess the effectiveness of the proposed ETAFHrNet model, we performed comparative analyses against multiple state-of-the-art segmentation approaches documented in existing studies. The QD-Crack dataset, along with several other publicly accessible crack segmentation benchmarks, was utilized to thoroughly evaluate the performance improvements and generalization capability of our model. For quantitative assessment, six evaluation metrics were adopted: Intersection over Union (IoU), mean IoU (mIoU), Precision, Recall, F1-score, Frames Per Second (FPS), and Params. These metrics collectively capture the model's segmentation accuracy, robustness, and inference efficiency. Specifically, IoU (Intersection over Union) measures the spatial correspondence between the predicted segmentation and the ground truth. It is calculated as the ratio of the area of overlap to the area of union between the predicted and actual regions. A higher IoU value reflects better segmentation performance. The metric is mathematically defined as:

$$\mathrm{IoU} = \frac{A \cap B}{A \cup B} \tag{10}$$

mIoU refers to the mean IoU across all classes and provides a comprehensive assessment of model performance.

Precision quantifies the proportion of true positive predictions among all samples predicted as positive, reflecting the reliability of positive classifications. It is formally expressed as:

$$\mathrm{Precision} = \frac{TP}{TP + FP} \tag{11}$$

Recall is the proportion of true positive samples that are correctly identified by the model, defined as:

$$\mathrm{Recall} = \frac{TP}{TP + FN} \tag{12}$$

Here, TP denotes the number of correctly identified crack pixels (true positives), while FP corresponds to background pixels erroneously classified as cracks (false positives). Conversely, FN represents crack pixels that the model failed to detect, incorrectly labeling them as background (false negatives).

Relying solely on individual metrics such as Precision or Recall can lead to a skewed evaluation, particularly when class imbalance is present. For instance, a model may achieve high Precision yet still perform poorly overall if Recall is substantially low. To mitigate this issue, we utilize the F1-score, which computes the harmonic mean of Precision and Recall, offering a more balanced and informative measure of performance in imbalanced scenarios. The F1-score is defined as:

$$\text{F1-score} = 2 \times \frac{\text{Precision} \times \text{Recall}}{\text{Precision} + \text{Recall}} \tag{13}$$

In addition, we introduce FPS (Frames Per Second) as a measure of inference efficiency. FPS quantifies how many input images the model can process and output per second. A higher FPS reflects a more efficient network capable of faster real-time crack detection, which is particularly valuable for practical deployment in infrastructure monitoring systems.

5. Results and Discussion

5.1. Influence of Semantic Labels and Transfer Learning on Model Performance

This section investigates the impact of semantic labeling granularity and transfer learning strategies on the performance of segmentation models. We evaluated four mainstream architectures—U-Net, DeepLabv3+, HRNet, and the proposed ETAFHrNet—across datasets annotated using both two-class and three-class schemes (see Table 2). The two-class scheme included only background and crack categories, while the three-class variant further distinguished between linear cracks and alligator (reticular) cracks. The findings indicate that models trained using the two-class scheme consistently surpass those trained with the three-class approach. For example, U-Net exhibited improvements of around 3.2% in mIoU and 4.13% in F1-score when utilizing the two-class dataset. DeepLabv3+ displayed greater robustness to label granularity, with metric fluctuations remaining within a 3% margin. Notably, ETAFHrNet achieved the best results under the two-class setting, reaching an mIoU of 74.41% and an F1-score of 83.11%. This superior performance can be attributed to ETAFHrNet's architectural design, which emphasizes long-range dependency modeling and directional feature enhancement. These mechanisms are especially effective when the task is simplified to binary segmentation, allowing the model to focus entirely on the structural continuity and contextual consistency of cracks without being distracted by inter-class ambiguity. In contrast, the three-class setting introduces greater intra-class variability and semantic overlap, which can interfere with feature representation and degrade performance. These findings suggest that the simplified two-class labeling strategy is more appropriate for practical crack detection tasks. Finer-grained class distinctions often introduce class imbalance and inter-class confusion [57,58], while offering limited added value in most real-world engineering applications.

We further evaluated the effect of transfer learning, specifically pre-training on a large-scale dataset followed by fine-tuning on a smaller, task-specific dataset, to assess its influence on model performance (see Table 3). The results demonstrate that this strategy leads to consistent and significant improvements across all evaluated models. Notably, ETAFHrNet achieved an 8.09% increase in mean Intersection over Union (mIoU), reaching a peak value of 74.4%, thereby outperforming all other models by a substantial margin. U-Net and SegFormer also benefited from transfer learning, with respective gains of 3.73%

and 5.63% in mIoU, further validating the effectiveness of this approach [59]. It is worth noting that Transformer-based architectures, particularly CNN-Transformer hybrid models, typically lack inherent spatial inductive biases and often require large-scale training data to achieve optimal performance. Consequently, pre-training plays a critical role in enabling these models to generalize effectively, especially when applied to smaller, domain-specific datasets, such as those used for pavement crack segmentation.

Table 2. Comparison of semantic segmentation model performance across multiple datasets.

Model	Classes	mIoU (%)	mRecall (%)	mPrecision (%)	F1-Score (%)
U-Net	two	62.85	70.76	71.23	71.90
	three	59.62	67.48	68.07	67.77
DeepLabv3_Plus	two	65.72	73.25	73.86	73.92
	three	62.47	70.09	70.54	70.31
HRNet	two	63.49	72.98	70.34	71.58
	three	60.15	69.11	67.32	68.20
ETAFHrNet	two	74.41	83.84	84.51	83.11
	three	71.08	79.21	76.64	79.42

Table 3. Results of transfer learning experiments.

Model	Transfer Learning	mIoU (%)	mRecall (%)	mPrecision (%)	F1-Score (%)
U-Net	No	59.12	66.32	67.28	67.45
	Yes	62.85	70.76	71.23	71.90
DeepLabv3_Plus	No	61.47	69.15	69.87	69.90
	Yes	65.72	73.25	73.86	73.92
HRNet	No	60.08	68.44	68.93	68.71
	Yes	63.49	72.98	70.34	71.58
SegFormer	No	63.58	71.72	72.13	72.25
	Yes	69.21	76.52	76.88	76.73
ETAFHrNet	No	66.32	74.65	75.15	75.38
	Yes	74.41	83.84	84.51	83.11

Based on the experimental findings, it is evident that both simplified semantic labeling and transfer learning substantially enhance segmentation performance, with ETAFHrNet consistently demonstrating the strongest results across evaluation metrics. Notably, the transfer learning setup involved pre-training on a composite dataset that included publicly available sources (e.g., CRACK500, GAPs384), before fine-tuning on QD-Crack. This configuration simulates cross-domain adaptation and indirectly reflects the model's ability to generalize beyond a localized dataset. To further investigate the critical factors influencing feature extraction and multi-scale fusion, and to conduct in-depth comparisons with alternative network architectures, we adopt the two-class labeling scheme and apply transfer learning as the default training strategy in all subsequent ablation and comparative experiments. This experimental setup is designed to ensure consistency and provide more reliable technical guidance for the deployment of segmentation models in practical pavement crack detection applications.

5.2. Ablation Experiment

To verify the synergistic contribution of the proposed CSHAM and EHAT modules to pavement crack segmentation performance, we conducted a series of systematic ablation experiments on a benchmark crack detection dataset. By progressively removing or replacing key architectural components, we quantitatively assessed the impact of each module on both segmentation accuracy and inference efficiency, measured in Frames Per Second (FPS).

As shown in Table 4 and Figure 9, the baseline model—comprising solely the original HRNet without integration of the CSHAM or EHAT modules—achieves an mIoU of 63.49%, with corresponding mPrecision and mRecall scores of 72.98% and 70.34%, respectively. The F1-score falls to 71.58%, and the inference speed is recorded at 14.51 FPS. These results indicate that conventional multi-scale fusion mechanisms, as employed in HRNet, are inadequate for capturing the elongated, fine-grained, and morphologically diverse structures characteristic of pavement cracks. Moreover, the visual outputs shown in Figure 10 reveal pronounced discontinuities and susceptibility to background noise in the baseline predictions, resulting in coarse segmentation contours and inconsistent structural delineation. These findings further underscore the importance of enhancing both feature fusion and directional awareness for high-precision crack segmentation.

Table 4. Ablation experiment performance comparison.

Model	EHAT	CSHAM	mIoU (%)	mPrecision (%)	mRecall (%)	F1-Score (%)	FPS	Params (M)
HRNet	No	No	63.49	72.98	70.34	71.58	14.51	45.0
	Yes	No	70.18	81.08	77.05	78.93	22.83	47.5
	No	Yes	72.69	79.95	80.96	81.45	22.64	48.2
	Yes	Yes	74.41	83.84	84.51	83.11	28.56	50.6

From a resource perspective, the baseline HRNet contains 45.0 M parameters. Adding EHAT or CSHAM alone keeps the footprint below 48.5 M while lifting mIoU by at least 7 percentage points and increasing throughput by 60%. Activating both modules brings the total to only 50.6 M parameters (+12%) yet almost doubles FPS (14.51 to 28.56) and raises mIoU by 10.9 percentage points, delivering the best accuracy–efficiency balance.

When the EHAT module is introduced independently, the model achieves an mIoU of 70.18%, with mPrecision and mRecall reaching 81.08% and 77.05%, respectively. The F1-score rises to 78.93%, and the inference speed improves to 22.83 FPS. As shown in Figure 10, the inclusion of axial positional encoding and local window attention enhances the model's ability to capture directionally oriented crack features, resulting in more continuous and clearly delineated crack contours.

In contrast, when only the CSHAM module is incorporated, the performance improves further, with an mIoU of 72.69%, mPrecision of 79.95%, and mRecall of 80.96%, yielding an F1-score of 81.45%. The inference speed remains comparably high at 22.64 FPS. As illustrated in the corresponding visualizations in Figure 10, the cross-scale attention and adaptive weighting mechanisms in CSHAM facilitate more effective integration of multi-resolution features. This enables improved detection of fine-grained crack structures, while preserving smooth and coherent segmentation boundaries.

When both the CSHAM and EHAT modules are enabled, the model achieves its best overall performance: an mIoU of 74.41%, mPrecision of 83.84%, mRecall of 84.51%, an F1-score of 83.11%, and an inference speed of 28.56 FPS. As shown in the rightmost column of Figure 10, crack patterns in examples (a)–(d) are accurately detected across multiple scales and orientations, with improved line continuity and significantly reduced background interference. In the regions highlighted by red boxes, the baseline and single-module variants exhibit noticeable segmentation gaps and discontinuities. In contrast, the combined use of CSHAM and EHAT yields contours that closely match the ground truth, demonstrating superior recognition of multi-directional and multi-scale cracks.

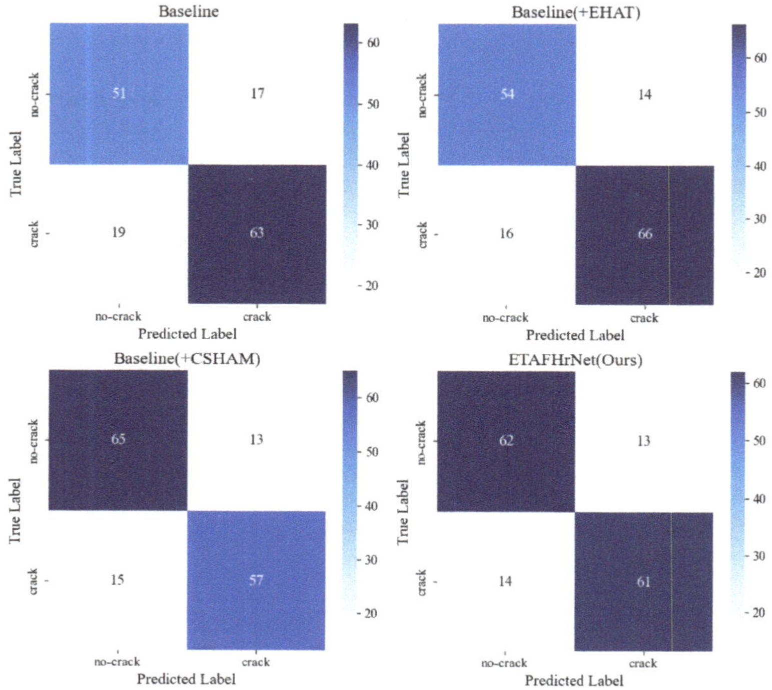

Figure 9. Confusion matrix visualization of ablation experiments.

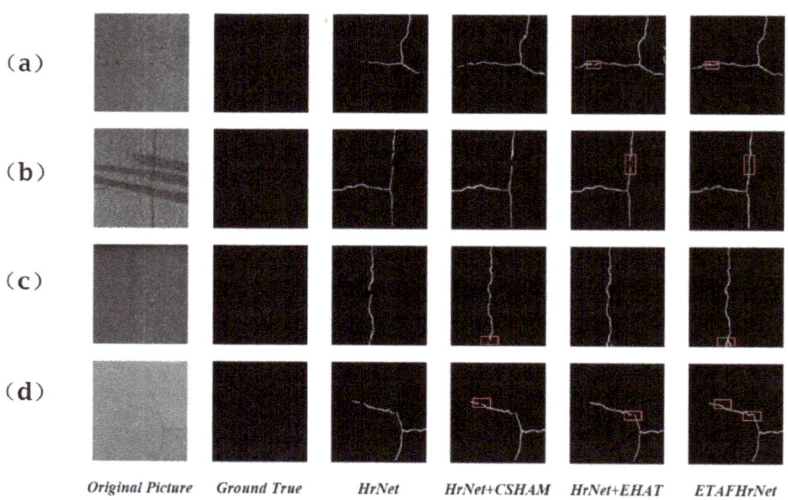

Figure 10. Ablation experiment comparison. (**a–d**) Randomly sampled images from the dataset. The red boxes indicate regions where differences occur.

In conclusion, the collective findings in Table 4 and Figure 10 underscore the synergistic contributions of the two proposed modules. The EHAT module primarily enhances the model's sensitivity to directional crack features, mitigating fragmentation and misclassification, while the CSHAM module improves feature expressiveness and background suppression through adaptive multi-scale fusion. Their integration leads to substantial gains in segmentation accuracy, structural consistency, and robustness under complex conditions. Additionally, the model achieves fast inference, rendering it highly applicable to real-time pavement crack detection scenarios.

5.3. Comparison with Existing Advanced Methods

To validate the performance advantages of the proposed ETAFHrNet model in pavement crack detection, we conducted comparative experiments on the self-constructed QD-Crack dataset against several state-of-the-art segmentation models, including U-Net, DeepLabv3+, SegFormer, PSPNet, and HRNet. All models were trained and fine-tuned under identical experimental conditions to ensure fair comparison. As shown in Figure 11, each model was evaluated using three primary metrics: mean Intersection over Union (mIoU), F1-score, and mean Recall (mRecall). The results demonstrate that ETAFHrNet achieves an mIoU of 74.41%, representing a 10.92% improvement over HRNet. In addition, it attains an mPrecision of 83.84%, an mRecall of 84.51%, and an F1-score of 83.11%, highlighting the model's significant advantage in segmentation accuracy and overall performance [60].

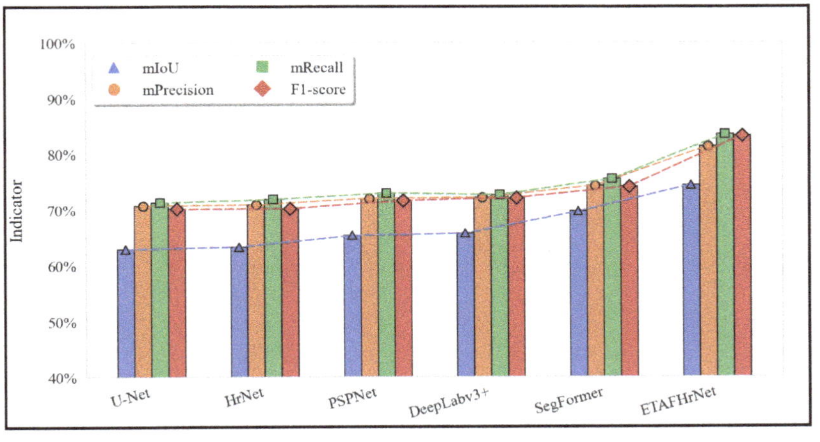

Figure 11. Performance comparison (mIoU, F1-score, etc.) of models on the dataset.

Besides accuracy, ETAFHrNet also offers a balanced hardware footprint. With 50.6 M parameters, it is only 12% larger than the HRNet baseline yet 21% smaller than the Transformer-based SegFormer (64.0 M). Despite this mid-range size, ETAFHrNet delivers an mIoU that is 10.9 percentage points higher than HRNet and 5.2 percentage points higher than SegFormer, while achieving the highest throughput (28.56 FPS). This accuracy-capacity-speed triad makes it attractive for edge GPUs and NPUs that typically provide 8–16 GB of RAM.

Furthermore, as presented in Table 5 and Figure 12, ETAFHrNet outperforms U-Net by 13.08% in mean Precision (mPrecision) and 13.28% in mean Recall (mRecall), indicating its superior overall segmentation performance. While DeepLabv3+ leverages atrous convolution to expand the receptive field, it exhibits limitations in modeling the continuity of slender cracks, resulting in a relatively low mIoU of 65.72%. SegFormer, constrained by its window partitioning strategy, tends to generate fragmented crack predictions during high-resolution detection tasks. Similarly, PSPNet, due to its coarse-grained context modeling, demonstrates a higher omission rate in fine crack detection, achieving an mRecall of only 72.64%.

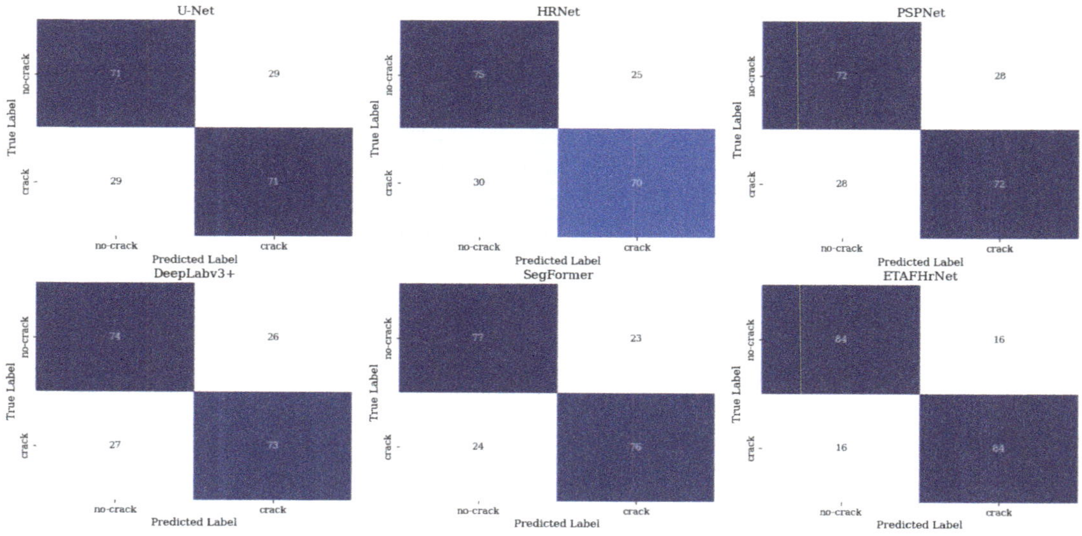

Figure 12. Comparison of classification performance across different semantic segmentation models based on their confusion matrices.

Table 5. Performance comparison of various semantic segmentation models on the QD-Crack dataset. (Bold text highlights better model parameters).

Model	mIoU (%)	mPrecision (%)	mRecall (%)	F1-Score (%)	FPS	Params (M)
U-Net	62.85	70.76	71.23	71.90	15.37	31.0
HRNet	63.49	72.98	70.34	71.58	14.51	45.0
PSPNet	64.32	71.85	72.64	72.24	17.22	42.6
DeepLabv3+	65.72	73.25	73.86	73.92	19.84	42.0
SegFormer	69.21	76.52	76.88	76.73	21.34	64.0
ETAFHrNet	**74.41**	**83.84**	**84.51**	**83.11**	**28.56**	50.6

Parameter-wise, all CNN baselines cluster between 31 M and 45 M, whereas SegFormer scales up to 64 M. ETAFHrNet falls between the two groups, indicating that its performance boost stems from architectural design rather than brute-force model scaling.

Given this footprint, we further estimate the real-time capacity of ETAFHrNet over a typical pavement section. Assuming one image covers roughly 1.5 m of pavement, analyzing a 1 km segment requires about 667 images. At 28.56 FPS, the model can process these images in 23.4 s (excluding I/O), confirming its suitability for near-real-time mobile inspection.

To provide a clearer illustration of the proposed model's performance benefits, we present a visual comparison using representative test samples, as shown in Figure 13. The figure presents original pavement images, ground-truth segmentation masks, and prediction results from ETAFHrNet, U-Net, and PSPNet, with red boxes marking key areas of discrepancy. The visual comparisons clearly show that ETAFHrNet offers superior performance in capturing directional and continuous crack features. In particular, for sample groups 1 and 4, ETAFHrNet accurately preserves crack continuity at junctions, where other models tend to produce fragmented outputs. In group 3, the model successfully detects faint, low-contrast cracks through adaptive multi-scale fusion, effectively mitigating the information loss seen in DeepLabv3+, which relies on a single-path fusion mechanism. In more visually complex backgrounds, such as those in groups 2 and 3, the integration of EHAT and CSHAM enhances both crack-to-background contrast and edge localization precision. By comparison, U-Net frequently exhibits crack discontinuities, attributed to its limited receptive field, while PSPNet often generates over-smoothed or mis-clustered predictions due to its coarse context modeling during feature fusion. These qualitative results further reinforce the quantitative superiority of ETAFHrNet in accurately segmenting diverse and challenging crack patterns.

Nevertheless, Figure 13 also reveals that ETAFHrNet is not flawless. (1) Sample a: the predicted transverse (horizontal) crack appears noticeably blurred compared with the sharper boundary produced by DeepLabv3+, indicating a tendency towards over-smoothing along horizontal orientations. (2) Sample c: the forked crack at the bottom is segmented with exaggerated width, resulting in an over-emphasized branch. These failure cases highlight the remaining optimization space for edge-preservation and scale-aware refinement.

Although the predicted segmentation maps from different models may appear visually similar in some cases, high-precision crack pattern identification plays a critical role in pavement management. Distinguishing between transverse and alligator cracks, for instance, informs whether surface sealing or full-depth patching is required. Precise segmentation also improves damage quantification, enabling more accurate cost estimation, lifecycle prediction, and prioritization of maintenance resources.

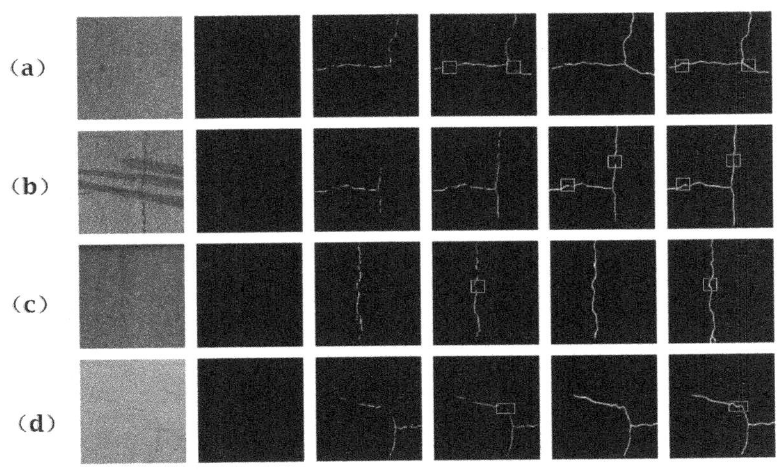

Figure 13. Segmentation outcomes comparison among different models. (**a–d**) Randomly sampled images from the dataset. The red boxes indicate regions where differences occur.

In summary, ETAFHrNet, empowered by its innovative hybrid attention mechanisms and adaptive multi-scale fusion strategy, delivers substantial improvements in segmentation accuracy, robustness, and computational efficiency for pavement crack detection. The model exhibits clear advantages in crack-structure preservation, background-noise suppression, and fine-detail restoration, underscoring its strong potential for deployment in real-world road-inspection and maintenance scenarios.

6. Conclusions

This study proposed ETAFHrNet, a Transformer-enhanced segmentation network specifically designed to tackle the challenges of detecting complex and irregular crack patterns in high-resolution pavement imagery. By integrating the Efficient Hybrid Attention Transformer (EHAT) and the Cross-Scale Hybrid Attention Module (CSHAM) into the HRNet backbone, our model effectively captures both long-range contextual dependencies and fine-grained structural features that are critical for accurate object segmentation and classification.

Comprehensive experiments on the self-constructed QD-Crack dataset confirm that ETAFHrNet surpasses state-of-the-art approaches, including U-Net, DeepLabv3+, and HRNet, in terms of segmentation accuracy, precision, recall, and inference speed. Ablation studies demonstrate that the two proposed attention modules provide complementary benefits, particularly in enhancing the representation of directionality, scale variation, and discontinuity, which are typical characteristics of asymmetric visual objects.

The proposed framework contributes to the development of interpretable and efficient AI models for infrastructure monitoring, with extensibility to a wide range of applications such as bridge inspection, tunnel lining analysis, and remote sensing-based structural assessment. Moreover, the model's architecture aligns with the broader goals of object detection and image classification, especially under challenging conditions where traditional models struggle.

Although ETAFHrNet shows promising segmentation accuracy and inference speed, several practical constraints remain: (1) Data diversity: the QD-Crack dataset mainly contains dry asphalt surfaces captured in daylight; performance under concrete pavements, wet conditions, night-time illumination, and extreme weather has not yet been validated. (2) Micro-crack sensitivity: hairline cracks narrower than two pixels are occasionally missed, revealing insufficient fine-scale feature capture. (3) Pavement-material dependence: preliminary trials on concrete surfaces reveal false positives where aggregate texture is confused with cracks, indicating the need for material-aware domain adaptation. (4) Edge deployment: the current model still relies on an NVIDIA RTX 3070 Ti GPU; additional pruning and quantization are required for real-time inference on low-power edge devices. (5) Continuous video streams: experiments were conducted on discrete images; real-time tracking of cracks in on-board video sequences demands further pipeline optimization. (6) Domain generalization: transferability to geographically distinct road networks or other infrastructure (e.g., bridges, airport runways) remains to be verified through cross-domain testing. Addressing these issues constitutes our immediate future work.

Looking ahead, future research will focus on optimizing ETAFHrNet for lightweight deployment on edge devices, enhancing its generalizability across diverse environmental scenarios, and improving its ability to identify micro-scale defects under varying pavement materials. More broadly, our findings emphasize the significance of modeling asymmetry and multi-scale variation in visual data, a principle that is critical for building robust, generalizable, and explainable object-recognition systems across real-world domains.

Author Contributions: Conceptualization, C.T. and R.L.; methodology, R.L. and Z.Z.; software, J.L.; validation, P.T., A.Y., and C.T.; formal analysis, C.T. and P.T.; investigation, S.P. and Z.Z.; resources, R.L. and Z.Z.; data curation, J.D.; writing original draft, C.T., J.L., and Z.Z.; writing review and editing, R.L. and P.T.; visualization, S.P.; supervision, R.L.; project administration, R.L.; funding acquisition, R.L. All authors have read and agreed to the published version of the manuscript.

Funding: This study was supported by the National Natural Science Foundation of China (Grant No. 42001414) and the "Elite Program" Research Support Foundation (Grant No. 0104060541613).

Institutional Review Board Statement: Not applicable.

Informed Consent Statement: Not applicable.

Data Availability Statement: For access to the data from this study, please contact the corresponding author.

Acknowledgments: The authors thank the anonymous reviewers for their constructive feedback, which greatly improved the quality of this manuscript.

Conflicts of Interest: The authors declare no conflicts of interest.

References

1. Huyan, J.; Li, W.; Tighe, S.; Xu, Z.; Zhai, J. CrackU-net: A Novel Deep Convolutional Neural Network for Pixelwise Pavement Crack Detection. *Struct. Control Health Monit.* **2020**, *27*, e2551. [CrossRef]
2. Ragnoli, A.; De Blasiis, M.R.; Di Benedetto, A. Pavement Distress Detection Methods: A Review. *Infrastructures* **2018**, *3*, 58. [CrossRef]
3. Oliveira, H.; Correia, P.L. Automatic Road Crack Detection and Characterization. *IEEE Trans. Intell. Transp. Syst.* **2013**, *14*, 155–168. [CrossRef]
4. Nafaa, S.; Essam, H.; Ashour, K.; Emad, D.; Mohamed, R.; Elhenawy, M.; Ashqar, H.I.; Hassan, A.A.; Alhadidi, T.I. Automated Pavement Cracks Detection and Classification Using Deep Learning. *arXiv* **2024**, arXiv:2406.07674. [CrossRef]
5. Mukherjee, R.; Iqbal, H.; Marzban, S.; Badar, A.; Brouns, T.; Gowda, S.; Arani, E.; Zonooz, B. AI Driven Road Maintenance Inspection. *arXiv* **2021**, arXiv:2106.02567. [CrossRef]
6. Li, Y.; Ma, R.; Liu, H.; Cheng, G. Real-Time High-Resolution Neural Network with Semantic Guidance for Crack Segmentation. *Autom. Constr.* **2023**, *156*, 105112. [CrossRef]
7. Liu, Y.; Yao, J.; Lu, X.; Xie, R.; Li, L. DeepCrack: A Deep Hierarchical Feature Learning Architecture for Crack Segmentation. *Neurocomputing* **2019**, *338*, 139–153. [CrossRef]
8. Sun, K.; Xiao, B.; Liu, D.; Wang, J. Deep High-Resolution Representation Learning for Human Pose Estimation. In Proceedings of the 2019 IEEE/CVF Conference on Computer Vision and Pattern Recognition (CVPR), Long Beach, CA, USA, 15–20 June 2019; pp. 5686–5696. [CrossRef]
9. Yang, F.; Zhang, L.; Yu, S.; Prokhorov, D.; Mei, X.; Ling, H. Feature Pyramid and Hierarchical Boosting Network for Pavement Crack Detection. *arXiv* **2019**, arXiv:1901.06340. [CrossRef]
10. Fan, R.; Bocus, M.J.; Zhu, Y.; Jiao, J.; Wang, L.; Ma, F.; Cheng, S.; Liu, M. Road Crack Detection Using Deep Convolutional Neural Network and Adaptive Thresholding. *arXiv* **2019**, arXiv:1904.08582. [CrossRef]
11. Chen, L.C.; Zhu, Y.; Papandreou, G.; Schroff, F.; Adam, H. Encoder-Decoder with Atrous Separable Convolution for Semantic Image Segmentation. In *Computer Vision—ECCV 2018*; Ferrari, V., Hebert, M., Sminchisescu, C., Weiss, Y., Eds.; Springer International Publishing: Cham, Switzerland , 2018; Volume 11211, pp. 833–851. [CrossRef]
12. Wang, X.; Girshick, R.; Gupta, A.; He, K. Non-Local Neural Networks. In Proceedings of the 2018 IEEE/CVF Conference on Computer Vision and Pattern Recognition, Salt Lake City, UT, USA, 18–23 June 2018; pp. 7794–7803. [CrossRef]
13. Hu, J.; Shen, L.; Sun, G. Squeeze-and-Excitation Networks. In Proceedings of the IEEE Conference on Computer Vision and Pattern Recognition (CVPR), Salt Lake City, UT, USA, 18–23 June 2018; pp. 7132–7141. [CrossRef]
14. Woo, S.; Park, J.; Lee, J.Y.; Kweon, I.S. CBAM: Convolutional Block Attention Module. *arXiv* **2018**, arXiv:1807.06521. [CrossRef]
15. Li, H.; Xiong, P.; An, J.; Wang, L. Pyramid Attention Network for Semantic Segmentation. *arXiv* **2018**, arXiv:1805.10180. [CrossRef]
16. Dosovitskiy, A.; Beyer, L.; Kolesnikov, A.; Weissenborn, D.; Zhai, X.; Unterthiner, T.; Dehghani, M.; Minderer, M.; Heigold, G.; Gelly, S.; et al. An Image Is Worth 16x16 Words: Transformers for Image Recognition at Scale. *arXiv* **2021**, arXiv:2010.11929. [CrossRef]
17. Liu, Z.; Lin, Y.; Cao, Y.; Hu, H.; Wei, Y.; Zhang, Z.; Lin, S.; Guo, B. Swin Transformer: Hierarchical Vision Transformer Using Shifted Windows. In Proceedings of the 2021 IEEE/CVF International Conference on Computer Vision (ICCV), Montreal, QC, Canada, 10–17 October 2021; pp. 9992–10002. [CrossRef]
18. Xie, E.; Wang, W.; Yu, Z.; Anandkumar, A.; Alvarez, J.M.; Luo, P. SegFormer: Simple and Efficient Design for Semantic Segmentation with Transformers. *arXiv* **2021**, arXiv:2105.15203. [CrossRef]
19. Liu, H.; Miao, X.; Mertz, C.; Xu, C.; Kong, H. CrackFormer: Transformer Network for Fine-Grained Crack Detection. In Proceedings of the 2021 IEEE/CVF International Conference on Computer Vision (ICCV), Montreal, QC, Canada, 10–17 October 2021; pp. 3763–3772. [CrossRef]
20. Fu, J.; Liu, J.; Tian, H.; Li, Y.; Bao, Y.; Fang, Z.; Lu, H. Dual Attention Network for Scene Segmentation. In Proceedings of the IEEE/CVF Conference on Computer Vision and Pattern Recognition (CVPR), Long Beach, CA, USA, 15–20 June 2019; pp. 3146–3154. [CrossRef]

21. Zheng, S.; Lu, J.; Zhao, H.; Zhu, X.; Luo, Z.; Wang, Y.; Fu, Y.; Feng, J.; Xiang, T.; Torr, P.H.; et al. Rethinking Semantic Segmentation from a Sequence-to-Sequence Perspective with Transformers. In Proceedings of the 2021 IEEE/CVF Conference on Computer Vision and Pattern Recognition (CVPR), Nashville, TN, USA, 20–25 June 2021; pp. 6877–6886. [CrossRef]
22. Ding, F. Crack Detection in Infrastructure Using Transfer Learning, Spatial Attention, and Genetic Algorithm Optimization. *arXiv* **2024**, arXiv:2411.17140. [CrossRef]
23. Huang, Y.; Shi, Z.; Wang, Z.; Wang, Z. Improved U-Net Based on Mixed Loss Function for Liver Medical Image Segmentation. *Laser Optoelectron. Prog.* **2020**, *57*, 221003. [CrossRef]
24. Zhang, L.; Yang, F.; Daniel Zhang, Y.; Zhu, Y.J. Road Crack Detection Using Deep Convolutional Neural Network. In Proceedings of the 2016 IEEE International Conference on Image Processing (ICIP), Phoenix, AZ, USA, 25–28 September 2016; pp. 3708–3712. [CrossRef]
25. Zamir, S.W.; Arora, A.; Khan, S.; Hayat, M.; Khan, F.S.; Yang, M.H. Restormer: Efficient Transformer for High-Resolution Image Restoration. In Proceedings of the 2022 IEEE/CVF Conference on Computer Vision and Pattern Recognition (CVPR), New Orleans, LA, USA, 18–24 June 2022; pp. 5718–5729. [CrossRef]
26. Chen, C.F.R.; Fan, Q.; Panda, R. CrossViT: Cross-Attention Multi-Scale Vision Transformer for Image Classification. In Proceedings of the 2021 IEEE/CVF International Conference on Computer Vision (ICCV), Montreal, QC, Canada, 10–17 October 2021; pp. 347–356. [CrossRef]
27. Tian, D.; Han, Y.; Liu, Y.; Li, J.; Zhang, P.; Liu, M. Hybrid Cross-Feature Interaction Attention Module for Object Detection in Intelligent Mobile Scenes. *Remote Sens.* **2023**, *15*, 4991. [CrossRef]
28. Wang, H.; Zhu, Y.; Green, B.; Adam, H.; Yuille, A.; Chen, L.C. Axial-DeepLab: Stand-Alone Axial-Attention for Panoptic Segmentation. In *Computer Vision—ECCV 2020*; Vedaldi, A., Bischof, H., Brox, T., Frahm, J.M., Eds.; Springer International Publishing: Cham, Switzerland, 2020; Volume 12349, pp. 108–126. [CrossRef]
29. Long, J.; Shelhamer, E.; Darrell, T. Fully Convolutional Networks for Semantic Segmentation. In Proceedings of the IEEE Conference on Computer Vision and Pattern Recognition (CVPR), Boston, MA, USA, 7–12 June 2015; pp. 3431–3440. [CrossRef]
30. Ronneberger, O.; Fischer, P.; Brox, T. U-Net: Convolutional Networks for Biomedical Image Segmentation. *arXiv* **2015**, arXiv:1505.04597. [CrossRef]
31. Zhao, H.; Shi, J.; Qi, X.; Wang, X.; Jia, J. Pyramid Scene Parsing Network. In Proceedings of the 2017 IEEE Conference on Computer Vision and Pattern Recognition (CVPR), Honolulu, HI, USA, 21–26 July 2017; pp. 6230–6239. [CrossRef]
32. Sun, K.; Zhao, Y.; Jiang, B.; Cheng, T.; Xiao, B.; Liu, D.; Mu, Y.; Wang, X.; Liu, W.; Wang, J. High-Resolution Representations for Labeling Pixels and Regions. *arXiv* **2019**, arXiv.1904.04514. [CrossRef]
33. Yin, Z.; Liang, K.; Ma, Z.; Guo, J. Duplex Contextual Relation Network For Polyp Segmentation. In Proceedings of the 2022 IEEE 19th International Symposium on Biomedical Imaging (ISBI), Kolkata, India, 28–31 March 2022; IEEE: Piscataway, NJ, USA, 2022; pp. 1–5. [CrossRef]
34. Xu, C.; Zhang, Q.; Mei, L.; Chang, X.; Ye, Z.; Wang, J.; Ye, L.; Yang, W. Cross-Attention-Guided Feature Alignment Network for Road Crack Detection. *ISPRS Int. J. Geo-Inf.* **2023**, *12*, 382. [CrossRef]
35. Lin, H.; Cheng, X.; Wu, X.; Yang, F.; Shen, D.; Wang, Z.; Song, Q.; Yuan, W. CAT: Cross Attention in Vision Transformer. *arXiv* **2021**, arXiv.2106.05786. [CrossRef]
36. Guo, F.; Liu, J.; Lv, C.; Yu, H. A Novel Transformer-Based Network with Attention Mechanism for Automatic Pavement Crack Detection. *Constr. Build. Mater.* **2023**, *391*, 131852. [CrossRef]
37. Li, C.; Fan, Z.; Chen, Y.; Sheng, W.; Wang, K.C.P. CrackCLF: Automatic Pavement Crack Detection Based on Closed-Loop Feedback. *IEEE Trans. Intell. Transp. Syst.* **2024**, *25*, 5965–5980. [CrossRef]
38. Chen, L.C.; Yang, Y.; Wang, J.; Xu, W.; Yuille, A.L. Attention to Scale: Scale-Aware Semantic Image Segmentation. In Proceedings of the 2016 IEEE Conference on Computer Vision and Pattern Recognition (CVPR), Las Vegas, NV, USA, 27–30 June 2016; pp. 3640–3649. [CrossRef]
39. Ho, J.; Kalchbrenner, N.; Weissenborn, D.; Salimans, T. Axial Attention in Multidimensional Transformers. *arXiv* **2019**, arXiv.1912.12180. [CrossRef]
40. Chen, Y.; Cheng, H.; Wang, H.; Liu, X.; Chen, F.; Li, F.; Zhang, X.; Wang, M. EAN: Edge-Aware Network for Image Manipulation Localization. *IEEE Trans. Circuits Syst. Video Technol.* **2025**, *35*, 1591–1601. [CrossRef]
41. Chen, J.; Lu, Y.; Yu, Q.; Luo, X.; Adeli, E.; Wang, Y.; Lu, L.; Yuille, A.L.; Zhou, Y. TransUNet: Transformers Make Strong Encoders for Medical Image Segmentation. *arXiv* **2021**, arXiv.2102.04306. [CrossRef]
42. Guo, J.; Han, K.; Wu, H.; Tang, Y.; Chen, X.; Wang, Y.; Xu, C. CMT: Convolutional Neural Networks Meet Vision Transformers. In Proceedings of the 2022 IEEE/CVF Conference on Computer Vision and Pattern Recognition (CVPR), New Orleans, LA, USA, 18–24 June 2022; pp. 12165–12175. [CrossRef]
43. Chen, Y.; Dai, X.; Chen, D.; Liu, M.; Dong, X.; Yuan, L.; Liu, Z. Mobile-Former: Bridging MobileNet and Transformer. *arXiv* **2021**, arXiv:2108.05895. [CrossRef]

44. Chen, Y.; Dai, X.; Chen, D.; Liu, M.; Dong, X.; Yuan, L.; Liu, Z. Mobile-Former: Bridging MobileNet and Transformer. In Proceedings of the 2022 IEEE/CVF Conference on Computer Vision and Pattern Recognition (CVPR), New Orleans, LA, USA, 18–24 June 2022; pp. 5260–5269. [CrossRef]
45. Wu, Z.; Liu, Z.; Lin, J.; Lin, Y.; Han, S. Lite Transformer with Long-Short Range Attention. *arXiv* **2020**, arXiv:2004.11886. [CrossRef]
46. Wang, Q.; Wu, B.; Zhu, P.; Li, P.; Zuo, W.; Hu, Q. ECA-Net: Efficient Channel Attention for Deep Convolutional Neural Networks. In Proceedings of the 2020 IEEE/CVF Conference on Computer Vision and Pattern Recognition (CVPR), Seattle, WA, USA, 13–19 June 2020; pp. 11531–11539. [CrossRef]
47. Zou, R.; Song, C.; Zhang, Z. The Devil Is in the Details: Window-Based Attention for Image Compression. In Proceedings of the 2022 IEEE/CVF Conference on Computer Vision and Pattern Recognition (CVPR), New Orleans, LA, USA, 18–24 June 2022; pp. 17471–17480. [CrossRef]
48. Vaswani, A.; Shazeer, N.; Parmar, N.; Uszkoreit, J.; Jones, L.; Gomez, A.N.; Kaiser, L.; Polosukhin, I. Attention Is All You Need. *arXiv* **2017**, arXiv:1706.03762. [CrossRef]
49. Tolstikhin, I.; Houlsby, N.; Kolesnikov, A.; Beyer, L.; Zhai, X.; Unterthiner, T.; Yung, J.; Steiner, A.; Keysers, D.; Uszkoreit, J.; et al. MLP-Mixer: An All-MLP Architecture for Vision. *arXiv* **2021**, arXiv:2105.01601. [CrossRef]
50. Shao, D.; Ren, L.; Ma, L. MSF-Net: A Lightweight Multi-Scale Feature Fusion Network for Skin Lesion Segmentation. *Biomedicines* **2023**, *11*, 1733. [CrossRef]
51. Jia, G.; Song, W.; Jia, D.; Zhu, H. Sample Generation of Semi-automatic Pavement Crack Labelling and Robustness in Detection of Pavement Diseases. *Electron. Lett.* **2019**, *55*, 1235–1238. [CrossRef]
52. Maguire, M.; Dorafshan, S.; Thomas, R.J. *SDNET2018: A Concrete Crack Image Dataset for Machine Learning Applications*; Utah State University: Logan, UT, USA, 2018. [CrossRef]
53. Zhuang, F.; Qi, Z.; Duan, K.; Xi, D.; Zhu, Y.; Zhu, H.; Xiong, H.; He, Q. A Comprehensive Survey on Transfer Learning. *arXiv* **2019**, arXiv:1911.02685. [CrossRef]
54. Russakovsky, O.; Deng, J.; Su, H.; Krause, J.; Satheesh, S.; Ma, S.; Huang, Z.; Karpathy, A.; Khosla, A.; Bernstein, M.; et al. ImageNet Large Scale Visual Recognition Challenge. *arXiv* **2014**, arXiv:1409.0575. [CrossRef]
55. Loshchilov, I.; Hutter, F. SGDR: Stochastic Gradient Descent with Warm Restarts. *arXiv* **2016**, arXiv:1608.03983. [CrossRef]
56. Kingma, D.P.; Ba, J. Adam: A Method for Stochastic Optimization. *arXiv* **2017**, arXiv:1412.6980. [CrossRef]
57. Jamal, M.A.; Brown, M.; Yang, M.H.; Wang, L.; Gong, B. Rethinking Class-Balanced Methods for Long-Tailed Visual Recognition From a Domain Adaptation Perspective. In Proceedings of the 2020 IEEE/CVF Conference on Computer Vision and Pattern Recognition (CVPR), Seattle, WA, USA, 13–19 June 2020; pp. 7607–7616. [CrossRef]
58. Cui, Y.; Jia, M.; Lin, T.Y.; Song, Y.; Belongie, S. Class-Balanced Loss Based on Effective Number of Samples. In Proceedings of the IEEE/CVF Conference on Computer Vision and Pattern Recognition (CVPR), Long Beach, CA, USA, 15–20 June 2019; pp. 9268–9277. [CrossRef]
59. Wu, B.; Xu, C.; Dai, X.; Wan, A.; Zhang, P.; Yan, Z.; Tomizuka, M.; Gonzalez, J.; Keutzer, K.; Vajda, P. Visual Transformers: Token-Based Image Representation and Processing for Computer Vision. *arXiv* **2020**, arXiv:2006.03677. [CrossRef]
60. Wang, Y.; Liu, C.; Fan, Y.; Niu, C.; Huang, W.; Pan, Y.; Li, J.; Wang, Y.; Li, J. A Multi-Modal Deep Learning Solution for Precise Pneumonia Diagnosis: The PneumoFusion-Net Model. *Front. Physiol.* **2025**, *16*, 1512835. [CrossRef]

MDPI AG
Grosspeteranlage 5
4052 Basel
Switzerland
Tel.: +41 61 683 77 34

Applied Sciences Editorial Office
E-mail: applsci@mdpi.com
www.mdpi.com/journal/applsci

www.ingramcontent.com/pod-product-compliance
Lightning Source LLC
Chambersburg PA
CBHW061624210926
51850CB00009B/2425
9783725855452